D1072680

Courting the Future

Courting the Future

preparing for a different world

Mark Ballabon

Eminent Productions Ltd
www.epl-uk.com

By the same author
Why is the Human on Earth?
A Second Chance at Life
A Short Journey of Poem Spirals

Companion Notebooks
The Chilled Zone
Sacred Spaces

Design and layout
Viv Mullett, Nick Ross

Original artwork and photography
Viv Mullett, Jackie Henshall, Julie Anne Glover, Mark Ballabon

With great appreciation to the editorial team of
Debbie Felber, Janet Donovan, Nick Ross, Viv Mullett,
Sophie Deen, Sarah Robins, Macarena Mata, Patricia Wallbank

Printed and bound in China

Paper from well-managed forests and other responsible sources

ISBN: 978-0-9559487-4-9

Published by Eminent Productions Ltd (EPL), Bath Place, Barnet, Herts EN5 5XE

Most of the concepts, frameworks and inspiration for this book come from sources of work copyrighted to Philosophical Frameworks Inc., and are used throughout with their kind permission. In acknowledging this, the author takes full responsibility for the accuracy of interpretation and usage.

Acknowledgement

There's saving someone's life by alerting them to danger or rescuing them. And there's saving and liberating someone's life, by demonstrating an example and leadership that inspires and guides a person to be free enough to be themselves.

And it's been the latter in the case of Leo and Ruth Armin (now passed away) and Ethra McKay, and what their lives have inspired in the author. Thank you for your vision, humanity, forbearance and the original, thrilling sparks of new perception that I have been able to use to light up my own life with a purpose greater than myself.

So this is my acknowledgement of their profound integrity and wisdom, and their mostly unseen and unrecognised gift of inspiring many other people around the world in the nurture and release of human brilliance.

* * *

Within all my endeavours, the constant encouragement from the lady I share my life with, has strengthened and moved me beyond my own limitations. The man woman mystery is indeed a very beautiful exploration of the unknown.

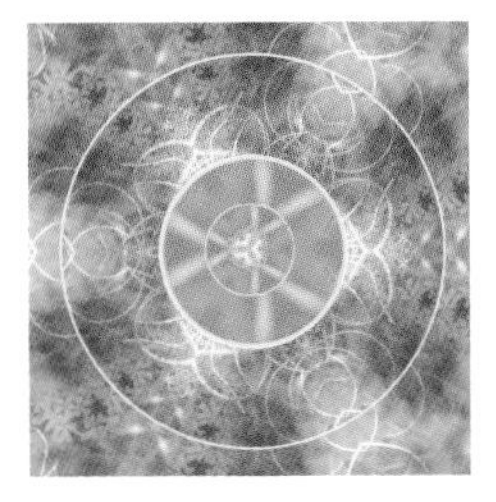

Contents

Liberating perception

A quest for life – meditations

the approach to anything
is everything

Welcome

There is a fine, intricate, amazing theatre of courtship to be wondered at in the planet's natural worlds, where dragonflies, seahorses, bowerbirds, deer and other creatures do the vibrant dance of attraction. In the case of human beings, we can court danger or hope, trouble or easement, prejudice or love. These are conscious choices that will determine what kind of a future we will attract.

Within this, there is a humility and courage to be found, which is a vital part of the process of courting; in which resistance and growth partner our every step forward.

The great adventures, new openings and extraordinary discoveries to be made in these times, come not from fighting or personalising or claiming the future, but by courting it. If we learn to court the future with the same passion and mindfulness with which we court a loved one, then we will embrace this emerging new epoch as partners in the great enterprise of evolution.

With such an approach we can become tomorrow's people today, no longer shackled by the hurts or triumphs of the past.

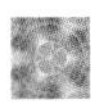

If the future is seeking anything from humans, it is seeking volunteers and pioneers who cherish the great value and wonder of living, who are open to what they don't yet know and are ready to change with the changing times.

And if we want a future beyond our time on earth, then perhaps the idea of courting the future becomes an essential attitude and position to take. For to court means to discover the ways of someone or something else – in this case the future – with the openness and wit to acquiesce and respond to needs greater than our own.

The children, and all those yet to be born into this mysterious and magical part of the universe, potentially carry the pollen of change and the new ways we may not as yet perceive. They ride the crest of evolution.

Will we also catch these incoming waves of change as they bring new intelligence, new power and new worlds of human possibility? Will we dive in without fear, overcoming the inevitable rip tides and undercurrents that come to sweep us off course?

Out here on the edge of an inconceivably vast galaxy – one amongst billions – we are a tiny, vital part of some very mighty purpose. Does the awe of this not somehow transport us into a different, timeless world, where everything is possible and the past carries no prevention?

In these rare, moving moments, there is a greater destiny to be touched and conceived, where such compelling and enhancing feelings as these join us to where the future is moving next.

The ingredients of this book were gathered from a lifelong quest to understand the causes and purposes of life from their deepest roots. This led to a range of new perceptions about issues that matter, and that could perhaps offer original insight into what our lives on earth propose. These are discoveries and practical methods that I really want to share.

From this endeavour has come a collection of essays, pictures and meditative writings, which seek out ways to attract, court and partner the future as it is appearing. Each writing is contexted within the inspiring, powerful and challenging times of change that we are living through now. For could it be that these human forms that we live in are in the flux of the most momentous evolution ever?

Whilst seeking to perceive and bring light to some of the greatest opportunities and concerns of current times, these writings also try to reach into generations that lie ahead, when the whole landscape, atmosphere and presence of life on earth will have irrevocably changed.

Such times are onsetting now, and are already testing our awareness, our readiness and our character of response.

My hope is that the pathways in this book tender an open invitation to that something inside that stirs as a deepest calling in each of us. A calling for self-realisation. A calling for service to humanity. A calling to venture into the unknown and the unexplored. Such is the quest to respond to the human proposal, and to what causes and gives life.

The direction of this book has been guided by the cruciality of seeking the right questions, for the frantic rush to solutions crowds out the kind of questions that can truly lead to solving and resolving issues. So the questions put forward are intended as meditations and practical exercises.

The essays themselves are offered as openers, starters, new viewpoints into important realms... a way in, not *the* way or *the* criterion.

Therefore the words in this book are written more in quicksilver than set in stone. For our understandings today will likely turn out to be merely glimpses from a much greater view to be perceived in the future.

Certain key themes permeate and reverberate throughout the whole book, like strong currents within the same flowing river: the themes of human perception, connection, freedom, letting go, vision, developing an open mind, exploring the unknown, the search for truth and dealing practically with the challenges and promise in daily living.

Each essay, picture and meditation that now follows is an interconnected part of an evolving and incomplete bigger overview. So it is possible to begin anywhere in the book that your instinct takes you...

Into the future

The future is not subject to time, as a first principle.

The future is the release
of original creation permission.
It brings new states, new intelligence, new hope.

A vision of the future

can you see one?

When someone is asked, ‘what is your vision of the future?’ and they reply, ‘world peace’, do they then meet with incredulity, or a cynical little snigger? Who is actually able to perceive and project a vision for the future that might one day achieve and go beyond world peace?

Holding and living a vision for the future depends on the level of perception in us. It depends on what we see the purpose for universal, human and planetary life to be. And it depends on whether we hold a vision based on an extension of the past, or on the fact of being alive, now, inside an infinite unexplored universe.

It's a question therefore of how broad a vision we can open our minds and intuitions to, and how real it actually is.

Whilst there is war, we dream of peace. Whilst there is starvation, we dream of enough food to satisfy all. Whilst the planet and her climate are under major stress and destruction, we dream of a sustainable future.

But a vision based solely on remedying something out of balance is a limited one – and that's often all we focus on these days.

Opening up a vision

How much do we think has been explored in terms of the human's ability to compose and play new music – like the music of the galaxies, music of the mentality, music to heal the skin? How many hundreds of sacred dances have yet to be felt, choreographed, celebrated? How many entirely new and original skills, crafts, conversations, designs or ideas still wait to be conceived in our minds and expressed?

Essentially, how much do we really know ourselves or what a human can be and do? Is it the same as what we know about the universe – maybe one percent or less?

There is a curious phenomenon in life, and it's summed up in this realisation that a friend who achieved an unexpected success told me about: 'No one said it was impossible. So even though I found out later that everyone thought I was crazy to even attempt it, I did attempt it, because in my mind it was always possible.'

So how many times have you thought you couldn't do something, and thereby proved yourself right?

Connecting to a vision of the future in the first place depends on one's ability not to prevent it, not to narrow it, curtail it, pre-empt it – as may be our habit. For we can often shrink our outlook on life and our view of what's possible down to our own personal experience, which is limited. Just think of how often you go through a day without ever looking up at the sky, or stopping to sense your own breathing, or sensitising yourself to the range and nature of feelings going on in you, or wondering where you go to when you fall asleep...

Vision is like the eagle that spreads its wings and soars to great heights because its view from the branch becomes limited. An open mind can do the same.

Whoever had the vision of placing a human on the moon's surface must have had to rise above all the concentrated minds in the world that believed it couldn't be done.

What vision your mind and brain see

It's revealing to search a little deeper here and look into two of the major faculties that open up vision in us beyond what we physically see: presence of mind and the creative processes of the brain.

We are born with an undeveloped mind, and a brain focused on survival. As we grow up, what we see in our 'mind's eye' depends on how we make up our mind, and then how well we've trained our brain to respond. For the brain can brilliantly process multiple thoughts in a second, whereas the mind can dwell, meditate and contemplate deep and complex matters that the brain cannot even conceive of.

For the mind is the main residence of human consciousness. It is thus potentially the highest human faculty in terms of its immeasurable capability, power and range for refinement, revelation and connection. A restricted mind however can soon become absent-minded, in two minds, mindless or even bloody-minded!

The mind naturally loves to discover the undiscovered, the mysteries, causes, reasons, and to develop its far-reaching scope in the search for truth. Our minds hold and develop the wisdoms we crystallise in living life. This is mindfulness or presence of mind.

The brain however is very much like a computer and processor, which, beyond automatic functions, is creative and responsive only according to what the mind programmes and influences and guides it to do, or not. Of course if a person has little presence of mind, it's likely that their brain will become a creature of habit and a camp-follower of the latest cultural opinions, trends and accepted norms. An overactive and dominant thinking brain seeks no higher authority than itself, and disperses mindful considerations in a person.

We are living through times in which we are rarely educated to develop mindfulness. This means that brains dominate: they over-control, design and shape the world we live in today. And a brain with little mindfulness seeks short-term results and ever more extreme stimuli to excite it and alleviate boredom.

This stunts the possible long-term vision that a human is able to realise, in which case the brain mainly thinks about the future as a reaction to, or a linear furtherance of, the past. This is one reason why people sometimes rationalise that there is no future to look forward to.

Over-brainy, mindless thinking creates a narrow, controlled vision of the future – a sterile world of serving robots, microchip implants in the brain, homes controlled by central computers, all communications mechanised and everything so high-tech, so comfortable, that maybe we won't even need to converse or feel or think much for ourselves anymore.

Many of the so-called advances in technology – from cars to planes to computers – are as unsustainable as the metal, chemical, oil, coal, electricity and gas supplies they run on; for if we project even fifty years into the future, we'll simply be running out of the resources to manufacture so many new products, or dispose of the old ones. So a vision for the future based solely on this kind of thinking doesn't have much of a future, especially with a population growing exponentially beyond seven billion people.

Human technology however – our mindfulness, thoughts, concepts, ideas, feelings, projections – is an unlimited resource to create with. Whether this is constructive or not depends on what we are aligned to and why. The choice, if we could see it, is stark.

For there are the sparking, networking, evolving realms of the future that are alive with the furtherance of originality, of genius, of love, of Creation permission, and there are the realms which afford the repetition of history and little more.

Humans being human...

... now there's a vision.

The sheer scope of connections, knowings, feelings and theatre that lies within the human capability is absolutely awesome.

Even after thousands of years in this updated Homo sapiens sapiens model in which we now live, the range of possibilities has hardly been explored. This may sound simple, but it's rarely ever a consideration when we plan for the future.

How much of a nation's vision of the future is rooted deeply within the greater development of inherent human talent, qualities and freedoms rather than narrowly within the growth of GDP and the proliferation of consumerism?

A vision of the future can be looking out through our eyes now, but such vision is blocked whilst we continue to see progress mainly in terms of material things external to ourselves. Are we more awe-inspired by the latest, most advanced gadget that we can hold in our hands, or by the wonder of what hands themselves can do, feel, touch, translate and sense, in this the most advanced of designs on earth?

If we only saw a vision of the future based on what our eyes can see physically, and what our brains have reference for now, then we would rule out the greatest part of any vision. For human vision is sight – but not eyesight alone.

Our senses, minds and feelings all 'see', by the same technology with which we see images and pictures in thoughts, dreams, meditations or moments of clairvoyance. This is much more far-ranging than the tight limitations within the mindset of 'seeing is believing'; for in the realms of a greater vision, seeing is receiving, seeing is conceiving, seeing is perceiving.

In which a human can mindfully develop the extraordinary technologies of intuition, extrasensory perception, conversation, attitude... discovering new ways to be natural with oneself and contribute to humanity. This is a matter of realisation in the fullest sense of the word.

So perhaps we need to look more deeply into ourselves, refresh our self-view, and develop more freely those amazing powers that we each have to explore what we can originally be and do as humans.

This may lead us to give to life more than we take from it – which is a great vision. And that can begin from this moment.

What is your vision of the future?

*

How can it begin?

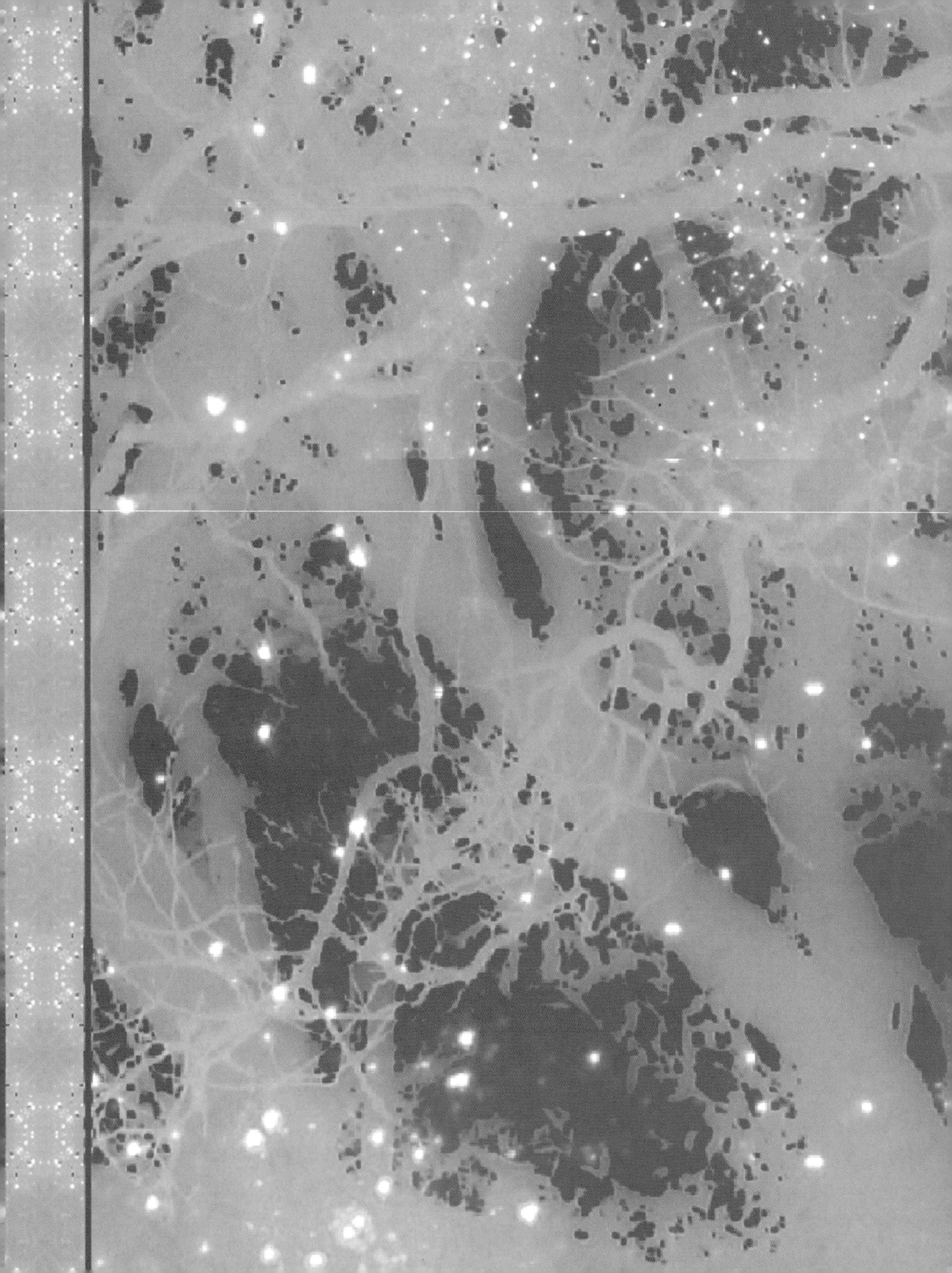

The edge of the unknown

into the inconceivable

The 25,000 mile journey around this planet could take two to three days, depending on which aircraft you fly in. Such distances and times we can conceive of. But what if we could travel to the nearest star, our sun…

Well, impossibly assuming we could fly the ninety-three million miles in a Boeing 747 and survive, it would take us about nineteen years of flight time to get to the sun. That's much harder to conceive of.

The measurement of light years is used to quantify the immense distances in the universe we inhabit, but how can we even imagine in our minds the nearly six trillion miles that just one light year equates to?

If we could take a trip to our nearest spiral galaxy neighbour, Andromeda, we would need to cover some two and a half million light years. That seems inconceivably far away from where we live. And today's most powerful telescopes can see distances of literally billions of light years deep into this universe.

Still, amazingly, we are able to look at the person sitting a few metres away and in the next second look skywards at the Big Dipper constellation, trillions and trillions of miles away in space.

How could words describe this wonder, let alone the fact that this great universe, in which we occupy such a tiny space, is still expanding?

And it goes further, deeper. For whilst there is this vastness of shining, pulsing, radiating life in the skies that we can observe, it 'only' makes up around 5% of the universe. New scientific discoveries indicate that over 70% of the universe is 'dark energy' and 20% is 'dark matter'. So how to even begin to think about what that means for us, living here on this relatively tiny speck floating around somewhere inside it all?

It's clear that in this universe and in life, there is infinitely more that is unknown than is known... including the immeasurable universe of our minds, our spirits, our souls and our emotions; this is like our own dark energy and dark matter, mostly unseen and unknown to us, or to anyone else.

How does it feel?

It is one thing to read about such vast dimensions and distances across space and time, but to actually get any real sense of this is something else...

What are the feelings that arise in the appreciation of such inconceivable realms of space, let alone the appreciation of the billions of unique moons, planets, constellations, nebulae, galaxies, lights, energies and influences that make up what we have so far discovered about this universe?

What nature of awe and amazement arises in us, and what feelings go beyond this? What of the knowing, haunting, intuitive feelings to be felt? And does this change the way we view and respond to living our lives here on planet Earth, as we spend another precious day inside this endless, expanding universal space and time?

Outlook

We each develop an outlook in life – it's our view of what is possible, however wide or narrow that may be.

Yet beyond our outlook on life, there's always what we cannot yet conceive of as being possible, in the same way that once human beings did not even conceive of farming land or inventing the wheel or using the digit zero. All these developments have taken place within the last 12,000 years – a fraction of time in the human story. So what could we conceive of *now* that has never happened on earth before?

In the last millennium many great discoverers and inventors have been derided, locked up or worse, as they faced the unforgiving opposition of minds closed beyond what they already knew and felt secure and comfortable with. How often do people react and fight against other people whose ways they can't or don't want to understand?

An outlook that is open, curious and seeking into the not yet known or discovered allows a far greater conception and vision of the future than one that is merely an extension of the already known. The difference is massive.

Into the inconceivable

It is possible to consider that which seems impossible for us to consider – however mysterious and contradictory that may sound!

In 1972 as an example, one scientist came to the view that 98% of human DNA was 'junk DNA'. This then became a widely accepted description and view. So instead of being open to consider the seemingly impossible – that DNA can actually do more than encode protein sequences – the very idea was simply dismissed.

Today however, whilst many of the functions of DNA are still largely unknown, most scientists reject the whole concept of 'junk DNA' because they now know it possesses not only biological functions but also great, undiscovered potential. So the mindset that would dismiss something as junk, because it cannot be understood or conceived of, can carry no vision.

Could it be that like the unknown universe, our unrealised and unreleased human capabilities and faculties, right down to the micro worlds of our very DNA, carry huge importance for the future?

Maybe they await the right timing, need, readiness and subsequent activation. In which case the evolution and development of the human, even as the progressive Homo sapiens sapiens that we now are, has only just begun.

A constructive approach to the future can begin in small, far-reaching realisations such as this.

Developing vision and trust in the future

Our vision for the future may be limited to a space-age high-tech paradise on the one hand, or a dramatic Armageddon on the other. Science fiction books and movies bring this vividly to life whilst often playing upon our greatest fears and hopes. But are such visions merely an extension of the world we view today and what we only conceive of based on current references?

Concrete, plastic and sophisticated machines may be today's tangible reality, but will they feature or even be needed in the future?

From the evidence of what we can see, understand and reason for ourselves, evolution and nature do not work along straight lines of progress. For who could have conceived of the radical development and change in the human itself, from the early basic Homo habilis model to the advanced Homo sapiens sapiens model that we now occupy? If we are to expect anything, it will be the unexpected.

For the future will bring more new events, more new opportunities, more surprises than we can even think of preparing for. Are we content then with holding in our open minds the future as a wondrous attractive mystery that we can acquiesce to, or do we feel the insecurity that drives us to fix and define a future centred solely around our ever-refining personal desires and comfort-seeking?

To conceive and construct a new vision, and to trust in the future, a mindfulness is needed that is unrestricted by habits, formulae and preconceptions, for this frees us to be able to respond to, and be responsible for, what the future itself may need.

So can we not trust in the future itself? Does it not always bring new life, new permission and new opportunity, irrespective of how humans may seek to taint it? Does the future not also extend to each human its trust, in freely giving every one of us the next moment to experience, even as you read this now?

If we could trust and work for it, tomorrow's world could be one in which progress and refinement will liberate and not harm; a world where our behaviours and decisions always consider the consequences for the generations to come – so that every child inherits an improving state of affairs on this earth.

This is not a dream world. It is a world to aspire to and endeavour towards, in which enlightenment and well-being would arise naturally from the dedicated human endeavour to be useful, honourable and of account to ourselves, to fellow humans, to this planet and the universe that sustains us.

In this spirit, there will be no fear of the future, only hope, value and eagerness to be with it.

The future is here to be conceived, to be courted, to be partnered for life.

What future can you conceive of,
unreferenced to these current times?

a layer of humidity

The passion of evolution

Evolution? How to even begin to open up possibly the most challenging, eagerly pursued, mysterious subject of all, that is so much more than a subject?

Do we trawl the millions of internet references, looking at endless pictures of skulls? Do we research all the 'isms' from Darwinism and Lamarckism to Creationism and other evolutionisms? Do we tread the well-worn paths that try to convince us that our ancestors were apes who gradually grew upright, became bipeds, managed to get a bigger brain, bypassed a few anomalies en route and ended up as Homo sapiens sapiens?

The study of the various developments of so-called hominids, or focusing on the line of Homo habilis, Homo erectus and why the Neanderthals never quite made it but the Homo sapiens line did, is only one approach. The study of skeletons and fossils is limiting, because evolution is a process that is not physical in its cause, only in its manifestation.

So where do we even begin?

Well, a first essential gateway is actually in the simple but crucial question, 'how do we approach evolution?'

Not as we would expect it to be, not as all the theories and hypotheses tell us it might be, not merely by collecting facts, but with all of our faculties and senses alert: with our minds, our mentalities, our feelings, our instincts, with our souls and our spirits, with every faculty that we possess as human beings, which includes the natural abilities of clairvoyance, telepathy and intuition.

Because in this way, we open ourselves to what we don't know, what hasn't been discovered, what is original, what we can sense and intuit and research in new ways that haven't even been dreamt of. And right there is our first big clue, *for these kinds of open, probing, detecting and uncovering ways are part of the very essence of evolution itself.*

And this approach takes us into a process of discovery first, not into a set of pre-fixed theories or the pressure to be over-definitive. For can the insistent pulse of evolution not be felt right inside the natural processes of growth, development, intelligence and generation that may exist from here to the furthest reaches of this universe and beyond – beyond black holes, wormholes, other universes, or anything that we can as yet imagine?

Thus we may draw a little closer to the way of evolution as it actually works; the way evolution itself reveals, unfolds, spirals, and pushes back the frontiers of the known universe, creating space for new growth.

So a natural way to approach evolution is from the understanding that everything that we are now, and are designed to be, has the character of evolution engraved into every single fibre of our being. *This makes us of evolution itself* and therefore able to detect, from within to without, what causes it to be.

For if not, it says that we are disconnected from the web of life and carry no meaning beyond an existence in a physical body – only able to treat evolution as an academic subject. Whereas the evidence we have with our own eyes, with every breath and new moment, overwhelmingly teaches us that we are part of everything that naturally exists and are designed to evolve with it.

Evolution is sparking around and through us all, as you read this now, inside the network of purposes of every natural living thing that stirs and has imprinted in it an urge to progress and develop.

Yet the human being is a unique case. Its evolution is not only marked by an inherent urge and will to grow and adapt, but by something else, unique to humankind – the urge to become more conscious, to refine. For humans, *this* process is not automatic, it must be elected into; whereas other natural life on this planet – from flora to fauna – evolves only by gradual automatic development and adaptation.

Therefore, if this first gate into the infinite domains of evolution is to open wide, then perhaps what's called for is a humility and open-mindedness that releases wisdom within a new discovery. In such a way, it may become possible for us to exercise and quest and enjoyably engage with the process of evolution itself, rather than just study it – respecting the volumes of written research, but not limited by them.

In this spirit, we can open up the versatility of evolution with a similar versatility that we possess as humans.

And language is an essential part of that human theatre. It is one of the core media through which evolution expresses itself and through which we communicate. Language evolves as we do.

The language of evolution

So what do we mean when we use the word evolution, and how do the two parts of the word, e-volution, actually form up?

This is where digging up the roots of a word can reveal a deeper meaning:

The letter **E** can represent – *energy, electric field, high force and power*

The suffix **-volution** can represent – *spiralling, unfolding, unfurling; patterns and processes that allow motion and change*

This leads to one possible definition of **E-volution**: *the force and power of change, through the natural patterns and movements of life.*

This takes us to another revealing approach, through a diagram:

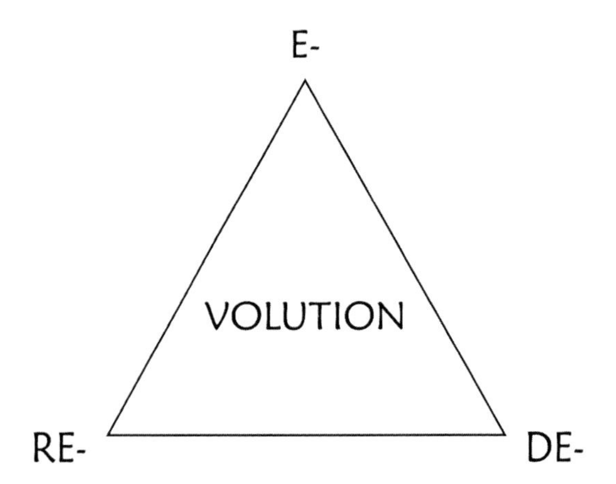

These three words all contain a linguistic root '-volution' which is to do with 'rolling' (Latin, *volvere*). With this in mind, the following clues emerge:

Evolution : unrolling, releasing, revealing, upgrading (Neutral)

Revolution : to roll back, pave the way, begin anew (Positive)

Devolution : to roll down, fall away, let go, open up (Negative)

We can find these three processes at play in the natural worlds, as we watch for example the caterpillar hatching from the egg, the chrysalis emerging as the caterpillar sheds its skin and finally the butterfly born free from its chrysalis.

At each stage in this life cycle there is a *revolution*, in which a new form emerges from the old, like the caterpillar from the egg. The caterpillar sheds its skin and the butterfly sheds its chrysalis – a process of falling away or *devolution*. Finally, the emerging butterfly constitutes a completely different form of life, a complete *evolution* from its previous stages.

Then the *revolution* of another life cycle and another and another – something of great beauty to behold.

Such processes go on automatically in nature, but evolution in the human case is directly linked to the development of consciousness and is a matter of our free will.

Evolution, in motion within us, is centred in the 'unseen worlds' of our mind, spirit, feelings and the constantly moving revolutions and devolutions of what we add to and subtract from our lives through experience. This has a physical manifestation in changes that take place every moment in our bodies and chemistry – from the makeup of our blood and genes, to the features on our faces.

This may offer a new and greater context of understanding, beyond history, and without any political inference:

Evolution as a neutral state, where new permissions for growth occur

Revolution as a positive force for change

Devolution as a negative force that facilitates the dissemination of change.

There is another perspective within this, in the curious harmonies of language.

From revolution to revelation, from evolution to elevation

There is a natural process of revolution, unrelated to wars, that goes on in terms of progressing away from an existing state of affairs, to reveal new perceptions, new ways ahead and new appreciations – such as during the Renaissance and the Age of Enlightenment.

We see this also in the 'spiritual revolutions' that seem to have driven the establishing of the earliest civilisations, rather than the classic view that it was agriculture alone. For it is widely believed that the settlement of nomads, and the beginning of farming around 11,000 years ago, were the primary reasons for the growth of villages and towns, with spiritual and religious movements arising later.

Yet there is clear evidence that a spiritual quest was also a prime motivation for early settlements, as can be witnessed in the extraordinary temple of Gobekli Tepe in Turkey, founded over 11,000 years ago.

With these kinds of spiritual revolutions can come the revelation and elevation of higher levels of personal and collective capability and development.

This in turn can trigger the process of evolution in a human's design, function and scope – especially in the realms of the mind and mentality.

So where does this take us?

This shows us firstly that language is alive, evolving, with depths of meaning beyond static definition. It shows us that language is an integral part of how humans develop intelligence. And it begins to take us further on the journey towards the essence of evolution itself, and further into the mystery and magic of the way of it.

To further understand evolution we need to be able to think it, feel it, touch it and realise at every step of the journey that we, as Homo sapiens sapiens – intended wise men and wise women – are actually a fine thread, a vibrant strand, an absolutely integral living part of the whole unending weave of evolution itself... unless we choose to disconnect ourselves from it.

Which takes us perhaps right to the threshold of a new understanding. For if evolution is connected to the power of consciousness and choice, gifted to us as humans with being, then it relates to what kind of a person we choose to be and become – either advancing the cause and endeavour of humanity, or making it into something lesser.

This is saying that our choices, and our self-determined development as people, can facilitate and join the very process of evolution itself.

In so doing, can we each not become a small, creative part of the flow of evolution that permeates the entire universe? For evolution is a great universal force, proposition and mystery; an internet of intelligence that applies to galaxies, space and something deep within the essence of every person.

An evolving story

Something extraordinarily mighty causes two powers, two massive forces, to inextricably come together inside a purpose: the pitching of Creation and the yawing of Creation, the ebb and flow of universal force and matter, the positive and the negative charges in an atom, a man and a woman.

And these two influences, when they meet and merge through opposite genders, cause a totally new appearance – a child. A new life. An original creation, seeded with the urge to refine itself.

And it is staggering, awesome beyond words, powerful beyond measure. It is a supreme passion, and that passion is the very pulse by which evolution unfolds its dynamic purposes and beauty by purposefully growing and refining life.

It is the passion of evolution, and when we find that our lives are moving on, are becoming of some account, are helping in some small way what life is here to do, we join this passion. It becomes ours too. And it bestows a deep feeling of fulfilment.

The universal influences, powers and intelligences that come through this passion have quickened the human in its own evolution, from the early model Homo habilis of two million years ago, to the refined Homo sapiens sapiens model that we are now graced with. And all the physical structures and parts of a human, right down to the individual cells and chromosomes, only change, upgrade and develop as an end result of such forces of nature from within and without.

We don't evolve as humans in a void. We are given the allowance and the consciousness – through reason and choice – to join and amplify the universal causes and purposes of evolution, or not.

In which your next new creative thought, compassionate act or original feeling may just be the next small step in your life becoming a more willing, active, happy participant in the evolutionary quest.

If this is the great proposal that life makes, do we consciously, willingly, want to learn how to play our best part in it?

What does evolution mean to your life?

*

What does evolution mean to humanity?

Language at a crossroads

There is a vast and growing vocabulary of over a quarter of a million unique words in English alone. And we know that Shakespeare used over 30,000 different words in his works. So why in our times do we mostly constrain ourselves to thinking, speaking and writing with just a few thousand?

Sometimes we can speak to each other as if we were texting or e-mailing, but using our voices instead of machines – or are we becoming too much like the machines around us?

Is it that in being swept up in the speed of modern life, we want to be too easily and quickly understood? And do we limit our depth of expression when we limit the depth and range of words that we make available to ourselves?

Sometimes writers, publishers or speakers edit out amazing words – like 'loquacious' or 'inchoate' or 'perfervid' – maybe because they think no one will understand them, or that they'll sound too 'posh'!

Our modern use of language seems to have been taken over by fast, short, clipped descriptions, in which the use of a wide range of available words continues to shrink.

Often, the almost unfathomable, deep well of available adjectives, verbs, metaphors, analogies and similes that are there 'at the ready' to describe ideas, concepts and the most subtle of feelings in glorious vibrant colours, gets squeezed and packaged into a murky shade of grey.

In many languages there are thousands of words that used to be thought, spoken and written in order to open up and finesse a great range of human observations, feelings and descriptions about the world that we live in – words now fallen into disuse. Meanwhile, other words seem to have had the essence vitamin sucked out of them by overuse or misuse, or have been replaced by new, invented, modern words which briefly pass in and out of fashion.

Why? Well here's an insight that may offer a route into the heart of this important consideration:

To not conceive of, think about or connect to something,
is to therefore not need or seek the words to express it.

For example, we know about and use the word 'metamorphosis' – which comes from the Greek meaning 'transformation' – because some people were able to conceive of, discover and articulate this process of transformational change that goes on for instance in the life cycle of the butterfly. So now the word 'metamorphosis' is a precious addition to our vocabulary and is ready, available and used today in a wider context.

But then consider, as one of thousands of examples, the noun 'equanimity'. It refers to mental and emotional balance, calmness, composure, aplomb. Some use it to imply a deeper sense of inner harmony, balance and conscious awareness. The word is therefore a subtle blend, describing a very fine state in a person, and is not often heard in conversation. So is it that finer or more honed words like this, which describe subtle states in humans, are rarely used today because we don't often feel these states? And thus the words describing such states have faded away through lack of use and need.

Has modern education and life experience become too formularised, over mechanised or habitual, to the point where we perceive and express ourselves across narrower territories, thus requiring narrower language use?

And has this contributed conversely to a burgeoning commercial, technical and specialist vocabulary, the detail of which can sometimes obfuscate the bigger picture and purpose of life, which requires much broader comprehension and description? We can name a galaxy M31, but can we name its character, purpose and influence?

So if we are no longer thinking or perceiving or exercising our minds across vast territories, we no longer employ or need a vast and diverse vocabulary.

But why does this matter? Well it depends on what you want.

Language is a crucial part of any personal or spiritual development, because its brilliant depth and range gives us greater depth and range to appreciate, understand and express ourselves, deal with others and engage with life more perceptively.

Language creates circuits and networks in the mind, mentality and brain, without which development is stunted. And the more accurate, rich, open and searching the language, the clearer, more engaging and meaningful one's expression can become.

This doesn't mean we should all swallow dictionaries. But if a person chooses, there is a mindset and attitude that can be developed that naturally draws on greater versatility with language, to access, appreciate and express the greater scope of experience possible in living life. A narrow mind will offer up narrow talk, whereas open minds can connect to the widest world of language and its manifold possibilities.

This is also not suggesting learning lots of long words – useful though that may be. What it highlights is that in living our lives, the broader and finer our toolkit of language, the more we can create and bring meaning to what we are and do. To convey a sentiment in a few well-chosen words is a fine art.

As an example, someone does something special towards your well-being and you tell them it was a 'nice' thing to do. But maybe there was a deeper sentiment, thoughtfulness and richness in their gesture towards you.

And so perhaps the description 'endearing', 'generous', 'considerate' or something similar would carry greater value and appreciation. Is it worth taking the care to find the language that fits the need of a particular situation, beyond generalisms, so as to convey more truly what we really feel or believe?

Of course integral to all language, is the way, emphasis and tone in which something is spoken, as well as the facial expression, body posture, inner feelings, reasonings and the look in the eye. These are essential parts of the great theatre of language, and ways to demonstrate to ourselves and others that we mean what we say.

In other words...

When a person drinks too much alcohol, not only do the central nervous system, basic functions and balance become impaired, but the 'higher brain' or frontal cortex, which handles conscious thought, becomes compromised. The culture of these times can drug us in exactly the same way, wherein the higher faculties of reasoning, feeling and perception can get anaesthetised or shut down by information overload, saturation and over-stimulation – leaving mostly an automatic and habitual reactivity to living.

When our senses are assaulted on all fronts, we tend to retreat into the apparent simplicity of sound-bite language which, like headline news, only reveals a tiny fragment of the true story – and this regularly gets mistaken for being the whole story.

So has the language we use become too based in expediency and economy, to the point where we are disconnecting from its natural complexity? And is this resulting in a meltdown of our vocabulary into more of a get-by exchange and mart?

Sometimes, when we feel insecure, we can end up speaking just to invite congratulations or confirmation of self. At other times, we may hide behind words and become political or deceptive. Yet all the while, the bounty of language is freely and cleanly available to us, to reveal, to think new thoughts, to be true to our word.

The great art of conversation is a marvellous developer of new and original understandings and ideas, whilst at the same time new words and phrases can be naturally and enjoyably discovered.

Do we mean what we say?

Whatever influences seem to be controlling and limiting language these days have clearly twisted many words away from the original – and often more natural – sense and meaning. And what now seems to define many English dictionary interpretations is common cultural, social or business usage of words.

One example of this is the word 'possess'. In most dictionaries the definition of 'possess' is to do merely with *owning* – as in the ownership of land – or with some form of demonic possession. But rare definitions relating back to earlier usage carry a completely different meaning for the verb 'to possess', such as (Oxford English Dictionary):

To maintain (oneself or one's mind or soul) in a state of patience or quiet.

And the following two quotations from letters dating back to the eighteenth and early twentieth centuries really show up today's limited usage:

A man who does not possess himself enough to hear disagreeable things without visible marks of anger... is at the mercy of every artful knave...
(Earl of Chesterfield, 1749)

Every man, worthy of the name, should know how to possess his soul – bearing with patience, those things which energy cannot change... (Mrs Lynn Linton, 1917)

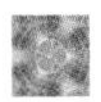

As another example, the word 'awful' used to mean 'striking or inspiring with awe or reverence'... so whatever happened to that?!

The point being that as we now live in such a materially-focused world 'out there', the language of the mysterious, the spiritual, the inner life, the unknown and the wondrous can become unexpressed or simply forgotten.

Does this then limit the development and evolution of human endeavour into the future?

For we may be able to easily access any word that we want off the internet in seconds, but can we access the vitamin, meaning and essence life of that word inside our own lives?

We become what we think about, and words can hone and enhance and elevate those thoughts and their effectiveness. Language is a code with endless permutations, like life itself, and different codes can open up doors into whole new realms of experience.

So when we speak, what is doing the talking? What is behind the rapid inter-exchange of almost robotic formulae such as, 'Hi, how's it going?' 'Not too bad thanks'? For these pre-digested formulae can lead to a 'whatever' kind of dialogue and a very limited experiencing of life. But that's a choice to be made.

What's important is that we are at least consciously aware that there is a choice beyond just accepting situations where our thinking, views and language are subservient to these formulae or social conformities or peer pressures. We don't have to just mouth words without saying what we really mean or without expressing the depth of what we really feel or believe, when the circumstance allows.

Where language itself is concerned, it openly awaits connection and application through the intention, enquiry and abilities of the user and responder. The option therefore always beckons, for us to discover and share some of language's greater riches.

For who could resist the amazingly kaleidoscopic language of love, of honour, of pride; or the language of the constellations, the subtle descriptive words for the hues to be seen in autumn, the fineness and nuances of human emotions, and the language of the profound mystery of purpose in how it all came to be in the first place?

For language is born of a need and an inner compulsion to feel, to discover, to know and to express and make sense of the lives we lead.

As these times change, so will language. Perhaps it will begin again, in a completely new phase in our and its evolution. Perhaps it will become less fixed to end results and more connected to processes, allowing nouns, verbs and adjectives to become more fluid, and new words to come into being arising from original thought rather than the abbreviation of existing words. There's certainly more of a future in that.

Exercising your mind in finding the words that fit is a fine ongoing development. For language can be very surprising when you allow it to be.

Imagine describing yourself, slowly, out loud, in order to really understand yourself in a much deeper and more conscious way.

Imagine, in doing this, that you reach beyond one-word lists – such as 'I am honest, impulsive, moody', etc. – and into a wide open, finely tuned searching.

Imagine touching, describing and expressing the deepest qualities that live inside you...

So how would you describe yourself in some depth,
now, as if for the first time?

*

How would you describe your purpose in life,
now, as if for the first time?

Questioning our way into the future

When we were very young, there was such a natural awe and wonder in the air; a presence, which each new wave of youth seems to be graced with. We were inside an amazing, timeless world of questions through which to discover life and its meanings.

This natural spirit of enquiry often gets suppressed and inhibited before puberty, as outside conformist pressures can shrink this eternal, compelling urge to deeply question everything.

For example, an adult asking a child, 'So what do you want to be when you grow up?' can force an oppressive position upon that young life, who at the age of say seven or eight can only be expected to snatch at an answer – possibly one which they think the adult wants to hear. Such leading questions pre-empt an answer and can restrict the faculty of questioning in a child's mind...

... whereas a whole kaleidoscope of new opportunities and self-knowledge can be released simply by listening to and encouraging the questions that naturally arise in children; the questions that maybe we once dared to ask and might now want to re-visit; questions that free our minds and senses to travel the unbounded possibilities of what it is to be alive, to be a human, to learn, to not know.

The right question at the right time can steer us to such fresh appreciations and understandings about life. Questions can not only liberate us from ignorance, but they can also give us access into greater ideas, thoughts and experiences. And questioning with the future in mind can lead to long-term resolutions, rather than short-lived, quick-fix solutions that don't dare to question at depth.

A question of 'why' and 'what', as a first principle

In approaching the search for questions, it's important to realise that the answer to a problem never comes from inside the problem. And the answer can certainly never be perceived from within the problem itself.

This is a trap fallen into so easily. We face a problem, we react to the problem, we try to fix it quickly. But when we do this, we tend to make our personal reaction to the problem – often heated and unreasoned – the basis upon which to solve things.

Such reactivity can cause us to instantly ask *how* to solve something, which immediately assumes we understand *why* it is happening and *what* we are dealing with. This may be fine in solving daily practical matters, such as how to get from A to B most quickly or how to dress a wound, but *how to?* is rarely the best approach to any complex issue.

How to solve an argument, conflict or crisis comes after asking why it occurred in the first place and finding the greater importances which lie beyond the dispute. Likewise, how to develop a quality in yourself, such as persistence or patience, comes after asking and finding out what the essence of these qualities actually is, and why you would want to develop them in yourself.

How questions are second principle, in that they naturally follow after the first principle questions of *why* and *what* have been explored.

In a time of economic recession, as an example, finding a way to understand the truth of the situation begins with why and what questions. Otherwise people can become quickly fixated on precipitous answers of how to solve it, such as 'print more money', which is merely a reaction to the problem from within the problem.

The net result in this case is to delay and worsen the problem, whilst temporarily sustaining the illusion of unlimited economic growth at any cost.

The heated thrust of modern living is towards immediate answers and solutions. Whereas in the key areas that concern the progress of humanity – from values and standards of living, to education and the environment – we seem to be getting too many answers wrong... especially when we aren't even looking for the right questions.

These tumultuous times of recession, depression and deep feelings of fear, are driving many people into a frantic rush for answers at any cost. For example, on the one hand experts are telling us that nuclear power will solve the energy crisis, whilst on the other hand we know how vulnerable nuclear plants are and that the legacy of nuclear waste for future generations is unconscionable. So isn't there a more crucial question to be asked before 'nuclear or no nuclear'?

We are also being told that stimulus packages and austerity measures will save the world's economies, whilst a soaring world population, unemployment and dwindling resources make this a nonsense.

So is it possible to pause, quietly, with the sanity and humility to say that *we first need to find the right questions to ask*? For without finding the right questions, we can end up only dealing with symptoms, side-effects and surface appearances. To not bother to find the right questions is to become lazily fixed by history, past experience and habit.

The kind of questions which will determine the long-term viability of the human race on this planet, are questions that are perhaps too simple and too searching for a heavily intellectualised and industrialised world, which sometimes seems only focused upon asking how to extract more from this earth so as to acquire more things.

So what of the questions that relate to the long-term aims and purposes that individuals and nations may be striving towards in this generation? And what will be the value and consequences of such aims over the next ten generations? Or are we questioning that far into the future? If not, how do we expect there to be one?

As we journey through this life, there are simple, revealing questions that can spark a renaissance of a natural questioning in oneself. This can develop character, understanding, tolerance and a greater sense of companionship and mutuality amongst people – a sense of true hope for the future. For certain fundamental, probing, illuminating questions apply to every person, no matter who they are – questions to reason through and quietly contemplate, such as:

What purpose does the universe serve?

Why is the human on earth?

Why is there such amazing diversity in life?

Why were the Pyramids, Stonehenge and the Easter Island statues built?

How do I see my purpose in life beyond what I want for myself?

What kind of a person am I trying to become?

What do I really consider to be a success in life?

What aspects of my character need improving?

What can I contribute with my life into this human race?

A clue in the word

The word 'question' has a fascinating and revealing construction of two parts – *quest* and *ion.*

The first part describes the natural and intuitive seeking inside every human being, which is a *quest.*

An *ion* is a charged particle, or energy with a purpose.

Therefore 'quest-ion' speaks of the quest for what is progressive, purposeful and constructive in living – what a person is charged to do by the fact of having been granted a life.

However, if we lose many of our natural connections and the passion and awe for living, then the freedom to really question can get squeezed into a narrow corridor of insecure self-interest and 'what's in it for me?' – as if this ever led to lasting happiness.

A question of the future

Is this not a crucial time then for humanity to find out what some of the most significant questions are – those pressing now, from the future, for a future?

There are pre-eminent questions to be heard and listened to that are naturally arising, often in young people. Some of these have been touched upon earlier and they promote further constructive, searching questions, which weave into networks of new intelligence and insight, such as:

What kind of greater purposes, beyond survival, exist for a human?

What is true self-leadership?

What kind of a planet will be left for future generations?

What is the future calling for from people?

What are the real reasons why humanity is facing the worst crises in climate, conflict, famine, poverty, economies, values, self-belief and any real sense of purpose to life?

Why are so many people, especially the youth, feeling a new sense of freedom and opportunity that carries no historical precedent?

Such questions can enable us to get a much better grip on the changes we are living through, and become more perceptive about the changes we want to make in ourselves.

If we really believe that children carry the seeds of the future, then we'll be carefully seeking for and responding to the questions that will directly affect tomorrow's world. Otherwise we may only be left with the kind of short-term questions that flit across the world stage, driven by a view of the future merely as an extension of the past.

Below, in the left column, are some of these oft-repeated, limiting questions, and on the right are some simple suggested alternatives:

What can I get out of life?	*What can I put into life?*
Why don't people understand me?	*How can I understand myself better?*
Why are there so many wars and crises in the world?	*Why is there such internal stress and conflict in people today?*
How to beat the recession?	*Exactly why are so many nations in a recession, and what are we learning from it?*
How to fight terror?	*Why does a war on terror cause more terror?*
How to stop global warming?	*What education is needed to change the behaviour that causes global warming?*
How to be rich and famous?	*Why is the human on earth?*

If we are to discover and understand some of the key mysteries and purposes in living a life, and apply the right answers to today's confronting challenges and extraordinary new opportunities, then we will need to be brave, patient and perceptive enough to find the right questions. There are many levels in this – from principled questions to practical ones – naturally so, because life is multiple, diverse and always moving on.

In the radically shifting and changing century we are living through, now is the most pivotal of all times to awaken and re-vitalise that crucial faculty that lives latent in each person's mind and spirit… the faculty and art of questioning, deeply. Our survival, our development and our future as a human race are all naturally constructed and determined by our ability to question. Questioning is a process that is not only about answers, but also about exercising our inherent freedom to discover the undiscovered.

What are the deeper searching questions
that you find yourself asking?

*

What do you think are the most important questions
facing humanity?

*

Why?

*

What is your own response to these questions?

The future in contrasting glimpses

Do we admit the future
its timeless breathing
into our mind space
to fly beyond our custom
to the birthplace
where we originate
new ways to be?

Do we watch its seasons
and their guiding touch
that shapes the signals
and moves the waves
of influence
which carry the reasons
to live on?

Are we still enough
to use our own clairvoyance?

Perhaps the charged light of future times, like sunlight, is too bright to look at directly, other than in glimpsed reflections on the water, so to speak. And so this analogy to do with the future suggested itself one morning, whilst swimming in and listening to the ocean tides...

If you walk to the shore of the ocean and pause before your feet step in, you can hear the world outside the ocean, as the waves pulse and flow against the shoreline, to and fro, in rhythm.

But when you walk into the ocean and immerse yourself completely in water close to the shore, you begin to hear a very different world. Its world in you.

You no longer hear the waves from a distance, but you feel connected to and swayed by the movement of the waves, as your ears fill with the delightful fine trilling of the shells as they are swept back and forth on the shallows of the ocean floor.

And yet, you have only entered an infinitesimal part of the whole ocean. You are only experiencing a tiny part of a tiny part of what the ocean is when it meets land, which tells you little of the ocean at depth, nor what moves the ocean or causes it to be.

This is a little picture of what it might be like for those now actively seeking to enter into and merge with this limitless ocean of a future. For the future is only 'out there' when we perceive it as a straight timeline. But it turns up 'in here', in us, in these times, when we feel its calling, its newness, its sense of everything being possible, and we become immersed in it.

This is to be witnessed especially through the eyes, and often extraordinary perceptions and questions, of the young; those who are still innocent to the taint of a modern world, and are still joined to the hope and promise of new life in a new epoch on earth.

For the young, the seashore can be a magical adventure. It is new, unexplored, dynamic, waiting to be discovered. And perhaps the young often feel the same about their lives, if they are not already prevented; for they begin in the fresh waters of the future, closer and more open to its nature and needs.

In this respect, the young generations are ambassadors of the future.

If true, why is this a significant awareness to carry?

People in their middle and later years are facing the next half or last part of their lives with the hopes, endeavours and concerns that come with say, forty, fifty or sixty years of experience of a life already lived. But a child of eight years old is facing a great unknown, with more to experience and more of a future to discover.

A boy or girl today is born into a time of huge changes in the human and planetary state of affairs. They grow up in a world of satellites, computers, mobile phones, hi-tech gadgets and many things that someone two hundred years ago never had and maybe never dreamt of.

The world of the young also includes an innate sense of hope and vision. For whatever gives humans life, imbues their starter years with this hope and vision as a deep motivation to explore what life can become. Yet alongside this, in these times, there is also an instinctive concern in the young – in some an urgent alarm – about what kind of a world is being left to them.

Their focus on the future is sharp, whilst for some adults it can be blurred, fading, or too habitual to even see the need to inspire a next generation.

Younger people carry the seeds and ways of the future from the new times they are born into. In essence, each younger generation is a new wave of 'hands on deck' for the ship that the future seeks to launch. But if adults do not set a wise and inspiring example, and leave the world an enhanced place to live in, then what has been the meaning and purpose of our lives?

The experience of life is there to teach us to know and act more wisely, to become an evolving human in the greater evolution of the whole species; this is the evidence of what has been happening on this planet for thousands of years. So are older generations a leading edge of wisdom and facilitation to the younger, or are they concerned too much with themselves? If it's the latter, then how will the future appear?

We are Homo sapiens sapiens, in which sapience or wisdom is meant to characterise what we are and what we can offer – else how do we expect this species to evolve?

Are we humans inside the future, or are we still carrying the weight and clutter of our past, and the human race's past? Have we become lost to the greater sense and vision of the times to come? For even in the first moments of a new day, if the debris of yesterday or last week or a problem from some years ago is our waking state of consciousness, then we prevent our ability to be in the next opportunity that the future is making.

The continual movements of sun and planet which cause day and night and make sleep possible, naturally grant us a cleansing of yesterday, to be primed for today and ready for the future. But when we awake, it is for us to elevate out from this natural period of rest, repair and preparation, and into a growing awareness of what the nature of the day brings, and what we can bring to it. Then we may be open and available to be influenced by the future and what it is making known to us today.

For the future carries the next opportunity to perceive, do something original, be wiser, move forward with belief in oneself and one's human pedigree. And when we overcome or let go of those episodes of anger or regret or frustration that take us away from being in the moment, we become more available to where life is unfolding, now.

Experience either liberates into greater and greater openness to what is beyond one's experience, or it becomes a rulebook of conformity from which a person can only do more of what they have already done. The past can become an indulgence, a bad habit, an obsession, or the rich fuel that inspires real progress and resists slippage.

To face the future is to feel your close self within it, only looking back for greater wisdom to make the next steps into the unknown.

Putting on one side, even for a few minutes, your personal experience of life, can you express anything about what the future actually feels like and might be calling for?

Letting go

Why worry?

Why am I so often pre-occupied? Why do I sometimes struggle with past hurts, future fears and my allergies to certain people and situations that can rattle around my head like old trains clattering across the railway tracks? Why do I worry too much? Do you?

Oh the sheer tonnage of bustling, clashing, inter-undermining, fragmented and heated thought processes which can pre-occupy our minds. Sometimes it's like carrying the noise of thousands of screaming fans at opposing ends of a football match inside your head!

Not all worries are damaging. Some understandable concerns can even lead to greater insight and intelligence. Worrying in a considered, constructive way about your health, your child or community may trigger all kinds of new ideas and helpful actions. But this essay addresses those kinds of temporary or deep-seated worries that are harmful, eroding us and our future from inside out.

So, how on earth is it possible to have a clear, cool run of self-chosen thoughts without the constant invasion, hijacking and insidious intervention of gnawing worries?

What reality looks like

There I am, awaiting an important call about a close friend who is very ill, and then the phone rings and a strange sales voice asks me disingenuously how I am today and do I want to insure my plumbing. What?!

This and a thousand other trivia, as well as ongoing problems, can weigh heavily upon us as we go about our lives in a world often over-absorbed by its ills, by its cult of status and by what isn't working. This can easily destabilise us and make us anxious and unsure of ourselves. Such an environment breeds worry in itself, in which any serious problem may compound the worry exponentially and distance us from ourselves.

When we are ongoingly worried, we make ourselves too vulnerable and too much of a willing and open marketplace for firms and individuals that want to sell us comfort food and drink, addictive drugs, plastic surgery, 'seven steps to bliss' self-help manuals and a multitude of products that promise to re-form a comfort zone around us. Well that may be fine in principle, but all this doesn't necessarily make the worries disappear, does it? So how to approach a way ahead?

Awareness

Awareness is half the remedy and relief, as well as being a great and organic development opportunity.

Not the awareness of your particular worry, problem or stress as a first principle, because if that is where you go to begin with, you will likely end up aggravating the situation. For when you find yourself only able to obsess about a worry or problem, it wears down your immunities and your presence of mind and renders you much less able to find resolution in anything. The answers are not to be found where the problems are.

Awareness can begin by locating and refreshing yourself first with respect to your aims in life, your values and importances – *not* with reference to what you are worrying about.

An important starter question in this process is 'What do you want?' For when you deeply search yourself with this question, you may begin to let go of what you don't want, and re-join what you really value and love about life.

Re-connecting to the flow and rhythm of life, and your life as a tiny engaged part of the web of all life, is the greatest healer and stabiliser to the rip tides that pull us under. And it provides a crucial, bigger picture awareness, perspective and context to any worry.

This ultimately enables you to identify it (the worry) rather than identify *with* it. If you identify *with* the worry and take it too personally, it shrinks your awareness and response down to the narrow confines of the worry itself.

Each day we spend on this earth can begin with awareness, even as we awaken and sensitise ourselves to a new day, open to the known and the unknown of what it may hold. In the overview and greater context that this allows, life's great mysteries and possibilities can govern our mindset, instead of giving in to the gnawing oppression that worries exact upon us.

Assessing the situation

Worries can retreat and be put into further context in the face of good reasoning that assesses the truth of a situation. This may offer a more intelligent and less dogmatic response to what's worrying us, as opposed to just yielding to habitual reactions of stress and panic. Worry can then become a leverage into knowing and doing better.

A nearby ticking bomb is definitely a worry, but trying to defuse it by rushing towards it and ripping off cables is not wise – unless you want to get your head blown off! You step away from the problem, assess the situation you are in, as you run to safety, alerting people to the danger. An expert can then be brought in, with the right tools, so that there is something greater than the problem to sort it with.

Similarly, an itchy skin rash is not sorted by scratching it. You try to discover the cause, which may be an internal disorder, and then research which creams, vitamins, exercises, meditations or whatever might work – always looking to improve your whole well-being as a first principle.

In a more subtle way, the grief of losing a loved one is intense, but will not be allayed by suppressing the grief, or trading on it, or self-pitying about it. Grief is a natural process of release which, given time and mindfulness, can enhance a greater understanding and value for the person who has died, and a greater appreciation for oneself and one's future. In such a way, after the initial sadness and heartache, grief can lead to an easement, a relief and a moving on, rather than an all-consuming worry, guilt or regret. Grief itself is not a problem, but our allergy or reactivity to it can easily make it one.

So you don't react to a problem, you find and respond to what the actual need is.

Finding a balance, before sorting the imbalance

When we are really worried, it's not a matter of going straight to the symptoms or problems or what is out of balance in us, for this can generate more worry and less understanding. One can't juggle the many balls that life pitches if one is standing on the edge of a cliff in a storm. It's a matter of beginning with the right inner location and balance from which to proceed.

Engendering a greater stillness and awareness within creates the quietness to listen to what's actually going on in you, and what it means. Such an inner awareness offers heightened perception and makes it easier to spot, understand and relegate worries.

For example, if a person wants to be more considered, calm, responsive, free and unworried in their behaviour, then a degree of ongoing 'inward self-commentary' during the day offers important guidance. This kind of talking, thinking and reflecting to oneself can become a powerful meditation practice, which in turn can neutralise worries. This inward dwelling is greatly aided by key questions, if they can be worked through with some time, space and minimal interference or noise. The ones opposite aim to provide some starters.

Opening, locating questions:

❊ What am I observing and feeling inside myself today, and why?

❊ What is most important to me about this?

❊ What in all these feelings is actually me, and what is being influenced by the nature of the day, the people around me and the current situation I find myself in? Can I separate out the differences?

❊ So what is in a state of good balance in me today, and why?

❊ What is out of sorts in me today, and why?

❊ If this is not what I want, what might help?

Next steps questions:

❊ Where do these insights leave me in the bigger picture of what I see to be my passions and purpose in life?

❊ Within this bigger picture, is there anything specific that am I aiming for today – and why?

❊ Is there a quality that I want to focus on practising today, such as honesty, better timing, bravery… ?

Working into such questionings as these, exercises the great gift of consciousness and choice, so as to lead our lives to best effect and prevent and release much of the stress and aggravation.

Choice

Yes it is a choice. The choice between what we will allow to pre-occupy and use us, and what we won't allow. It's clear for example that going shopping, out on a date, to a meeting or to work can be done with anxiety, apprehension and impatience, or with self-determined ease, adventure and a search for new opportunity.

Whether we find ourselves in challenging or comfortable situations, we choose how we want to be, either deliberately or by default. So there is a need to be as conscious as possible in this, because otherwise something or someone else is making the choices without our agreement – which is worrying! This may all seem obvious, but exercising choice works or breaks down at so many levels – subtle and unsubtle…

Look how easy it is when someone close tells you of a problem they are having with a mutual friend, and then without hearing the other side of the story, you immediately become biased by and pre-occupied with their worries, their outrage and their version.

Choosing not to take sides, choosing to be more neutral in times of conflict – especially inner conflict – choosing to listen and look for what hasn't been discovered or understood yet, is a means to be pre-occupied in the most effective way. And it leads to an ongoing personal development with a lot less anxiety.

A contemplation along the way…

You are what you accept, what you reject, and what you connect to

We each have the freedom to make conscious decisions about what we will accept or reject. In the absence of this, anything and anyone can contol, dominate and disconnect us from the harmonic rhythms and course of life.

One key development issue in this is to knowingly build the natural standards and principles and values by which we want to live. This is best done as an ongoing exercise, contemplation and working out – either called over to oneself, expressed to a friend and/or written down and regularly reviewed.

It is in these ways and with these growing connections that we come to know ourselves better. And feeling such a secure foundation inside oneself acts as a great easement in times of worry, crisis or tragedy.

Preventing stress and aggravation

Seeking to prevent stress and aggravation is an excellent ongoing, guiding principle in life – an oasis that fends off worry.

Fortunately, there are so many ways to do this.

We can always find ways to treat ourselves and others more kindly; ways to take ourselves less seriously; ways to be less hard on ourselves; ways to slow down when the pace of living makes us go too fast for our own good.

We can also surprise, relax and release ourselves in the spontaneity of getting up and dancing a new dance, singing a new song, writing down a new idea, exploring or attempting something we've never tried before. We may then find that the anxieties loosen their depressing hold.

Worries will come and go, and some may stay, and this we simply need to understand about the way that life is today. However, the steering wheel that turns us onto better routes and away from troubles is always in our own hands – providing we can keep a grip on it.

When all else fails, a good sense of humour never does!

Fortunately, we are not the centre of the universe, and the planet revolves quite happily around the sun without throwing any tantrums or worrying itself to death.

When nothing satisfies

So many things to do and be in life. So much catches our attention. So much to want, so many places to see, people to get to know, things to be inspired by. All these chances to be happy and satisfied – but we're not that often... are we?

It's interesting to read the league tables – for example in the University of Michigan's *World Values Surveys* – listing which nations are said to be happier than others. But even the supposedly happiest of nations seem to have a high percentage of very miserable people!

Here is a perception that may help to contemplate this further:

Happiness is escaping that which prevents you.

So what might reveal some of the inside story about what prevents a natural flow of satisfaction and happiness?

The oldest part of our brain, centrally located in and around the brain stem, controls essential survival and maintenance functions as well as basic behaviour. Its workings and range are much more rudimentary than the two other major parts of our brain – limbic and the cerebral cortex – and it exercises a lot of control. This works brilliantly when it comes to breathing, balance, circulation, digestion and so on, but in terms of our behaviour we can allow it to exercise far too much control…

… for when we meet different situations every day, especially challenging and confronting ones, this primitive part of our brain can instantly cut in and override any higher reasoned, mindful approach which works through the more conscious part of our brain, the cerebral cortex. And without the balance, calmness and sensitivity that a developed mindfulness would bring, the primitive brain tends to be rashly reactive rather than wisely responsive.

Does this suggest that we have given this basic part of our brain too much dominion in our lives, and extended and distorted what it's designed to do? Has this in turn led to excessive control, management, competition and authority in human behaviour? And is this why the left hemisphere of the brain – which tends to process such actions – has become over-dominant at the expense of the right hemisphere which processes our more creative and empathic feelings? Perhaps all this has led to us becoming too constrained and governed by our basic female/male reactions, rather than becoming freer in our improving woman/man responses.

In these times of instant, pressured decision-making, the primeval region inside our brains can take on a rapacious, territorial and self-centred outlook, causing us to covet and hunt things and people and situations that we think we want – and then seek ways to keep them.

But then a strange thing happens. *If we get what we want, we are not satisfied because we want more. If we don't get what we want, we are not satisfied because we didn't get it.*

So often this adds up to rarely being satisfied with anything.

There is a profound illusion going on in all this. And the illusion is that satisfaction should be our main goal in life. But the irony is that if satisfaction is our main goal, it will be virtually impossible to ever achieve.

It's like the mythological story of Tantalus (whence the word 'tantalise'), in that the more we reach for and try to possess and control the important things in life, the more they tend to retreat and become elusive.

We can't just *make* it happen all the time. Often we need to *let* it happen. The seeking of love or respect or esteem from another person cannot be forced or pressed, for such qualities begin by being consciously developed inside oneself. This then becomes satisfying in itself, without needing to clamour for the approval of others.

The word 'satisfy' is from the Latin *satis*, meaning enough, and *facere* meaning make – thus to make or have enough. And this underlines one of the challenges in being satisfied…

In the industrialised world, we are heavily persuaded into a consumerist and commodity-based attitude that has a background noise of nothing-ever-being-quite-good-enough. And why? Is it because someone else always seems to have more of what we want? Or is it also because we are instinctively seeking something much deeper than status, fame or material wealth will ever satisfy?

We are mostly conditioned to be very competitive and comparative creatures, evaluating our self-worth purely in relation to everyone else, and rarely in relation to aims, standards and principles that we ourselves decide and set.

Then comes the fact that our expectations nowadays can be so unreasonably high that nothing is going to truly satisfy. And whilst we may instinctively know this, unrealistic expectations are constantly being fuelled by poor reasoning, insecurities, fears and drugs of many kinds – all of which tend to stymie the throughout satisfaction of knowing oneself better.

Notwithstanding all the disappointments, hurts and sadnesses that everyone goes through as part of life's journey, there is still a magical inner satisfaction and peace that is always achievable. In extreme situations of poverty, starvation or violence this may prove unattainable, but the possibility itself never retreats, and is there if we can reach it.

That inner satisfaction, when it is touched, is one of a profound, simple irrepressible contentedness; not a self-satisfying or smug contentedness, nor even just a satisfaction with one's lot. It is to be felt in the sheer awe and value of finding yourself alive, each day, on a planet of great beauty and beneficence, with an absolutely individual chance at life.

Such warmth of gratitude is a self-determined attitude.

For the fact of being given a chance at life is supreme. Anything else is a great blessing upon an already very great blessing.

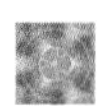

Satisfaction, in its highest reaches, is a natural enhancement that overflows in joining and working for a purpose greater than oneself – to cherish, to enjoy, to serve. This may be felt, even in a passing moment, when we become aware of ourselves as a small, powerful, responsive part of something infinitely greater, vaster, caring, unfathomable, destinic.

Such deep satisfaction also rises up inside when we feel we can contribute and usefully volunteer into a greater cause. This happens in many different ways around the world – often going unnoticed or unrecognised – in simple acts of kindness for example, especially when the sentiments are strong and reasoned and are not just a one-off gesture. This in itself can take many forms, from creating new opportunities for others and alleviating stress and suffering, to being a quiet, steady, peaceful presence and support in times of conflict.

Such actions and motives also inspire a sense of belonging, which is deeply satisfying to that natural compulsion in us to become part of a great enterprise and to do something useful, lasting – even more so when we share and work this with others. For camaraderie also satisfies an inherent human desire to share interests and beliefs and ideals with others of kind.

Curiously, being dissatisfied is a good opportunity to review our expectations and values and what we see to be the real importances in life; an opportunity to re-centre our lives and let go of mental clutter. And it can lead to a greater perspective, balance and contentedness which consequently ensures that unimportant things remain unimportant.

Understanding what we really need and want in life is a great development journey. For to want more than you need never satisfies.

By contrast, spontaneously volunteering the right response at the right time to a real need can offer an unexpected and thrilling sense of fulfilment.

So is it that we are satisfied, throughout, when we seek beyond self-satisfaction?

Perhaps it is when we are generous with our time, our talents and what we love about life, that we can really feel at ease and clean inside our own skin and at peace inside our own mind, never feeling lesser or greater than anyone else.

Is this not a freedom from the forced pressure to satisfy and be satisfied?

What do you find truly satisfying?

*

Why?

Without prejudice

re-finding an open mind

Whatever reasons cause human life to be, whatever our experience in the womb and whatever our genetic makeup, we didn't inherit prejudice. From our first moments on earth however, we were swamped with it.

Much of the environment that a new life is born into these days carries prejudices arising from all sorts of dogma, unyielding moralities, taboos and the crude or subtle pressures to behave in a certain pre-determined way. This is a significant part of the formatory world that strongly influences so many young lives, whether it be through their family, other people, home life, schooling, national conformities or the whole lot. It affects us all.

In natural circumstances, if a child's education followed their innate growth patterns and their unique character and needs, they would gradually learn how to discover and self-determine their own way in life. But if every step that they try to take as they grow up is being forced towards following in someone else's footsteps, someone else's beliefs and opinions, or someone else's pressure to conform to what they think is the only way, then a person's progress in life can only become prejudiced.

Prejudice is like a latent or active pathogenic virus that we catch in our youth and carry in our systems from then onwards. And when it flares up, in anger or frustration or weakness, it exercises an intolerant, ignorant and oppressive bias. As such, prejudice can only harm.

Prejudgement however is different, for prejudgements are not always damaging. We may for example prejudge what a new situation will be like before we have all the information, simply because we want to prepare ourselves or ensure our safety as best we can beforehand.

Prejudging what an up-and-coming interview or imminent trip overseas will be like, or prejudging best and worst case scenarios, may usefully help in your preparations; in other words, you may just be trying to assess wisely and objectively before such circumstances occur.

Likewise, the cover of a book, the packaging of a product, or the goods in a shop window may all bring out prejudgements in us of one kind or another, for good reason or not. Yet mostly this is as harmless as prejudging that it will rain because the storm clouds have gathered, or prejudging that cheap food or cheap items will be of lesser quality because of previous experience – it's reasonable, whether right or wrong.

All these kinds of prejudgements can be a useful or expendable part of living and surviving in today's world, where we may often prejudge without doing any harm. For even if the things that we prejudge actually turn out to be completely different from what we thought, it's unlikely that anyone got hurt and we'll have learned something from the experience. So prejudging and making informed assessments beforehand is, in essence, neutral.

However, when we prejudge ourselves or another person or a situation, with a preconceived opinion that attacks or demeans or dismisses any other view, then it turns into outright prejudice.

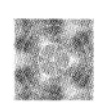

Prejudices are learned very early on in life, or in some cases are deliberately indoctrinated into us at an age when our brains and feelings are highly impressionable. This can breed an unforgiving and aggressive approach to life, perhaps masking the weakness of character that prejudice instils.

Our ignorance, insecurities and unreasoned fears tend to thrive on prejudice, because prejudice offers the apparent security of having a fixed opinion, having something definite to say, believe in and identify with. But that's a false security, often based on our unreasoned reactivity to the people, situations or things we like or dislike.

And of course prejudice often leads to instant and damning judgements upon people. So we can end up for example having all sorts of prejudice flooding into our heads when we first meet someone and instantly judge them by their accent, colour, clothes, face, shape, hairstyle, smell, walk. And we may do this without even knowing a single thing about that person. Similarly, we may quickly become prejudiced against someone who looks at us with apparent anger, sharpness or mistrust, and decide to avoid or even antagonise them.

We can also end up with all kinds of prejudices about major issues in the world whenever we take sides based merely on popular or soundbite media opinion. Or we may become prejudiced against someone else based on rumour, one side of a story, or because maybe we just resent their success.

Such forms of prejudice taint so much of life and what the future holds. They pre-empt, pre-fix and pre-determine an outcome before time and before there's enough unbiased information to come to an accurate viewpoint. Most prejudice is therefore an inability or refusal to perceive the truth of things, with the belief that our personal view is the only one that counts. Prejudice can thus lock us into a narrow world of self-deception, restricting our journey in life as spiritual beings seeking human experience, and trapping us in a culture of inhumanity.

This also highlights one of the greatest prejudices in this world, which is actually the prejudice of a person against themselves.

This begins when we fix ourselves, and our self-view, inside the narrow confines of our experience of life so far – good or bad – which roots us in the personal history of what we were, not in what we can become, now, for the future. It may also cause us to box other people into our limited experience of them, blinding us to their essential nature, their unspoken life story, their deeper motivations and the changes that may be happening in them.

Prejudice against oneself can become a denial and a prevention of the natural capabilities, passions and pedigree of a human being. It is a root source of the prejudice that is exacted upon others, which is essentially an outlet for trying to rid oneself of the very oppression, blame, bad memories and toxins that accumulate when we judge ourselves harshly, or were similarly judged by others. For the simple truth is, we tend to hurt others, wittingly or not, when we are hurting inside.

You don't need prejudice to observe and assess and deal with a person, because this narrows and obstructs your ability to see them as they truly are, stifling respect and value.

And when a person is prejudiced against another person or group, they are also prejudiced against humanity, for that other person or group is a part of humanity. This is a prejudice against oneself too, for we also are a part of humanity.

Meanwhile, through it all, the mind longs to be free to exercise itself, without the shackles of ignorance that prejudice binds us with; the mind longs to assess, appreciate and be open to what it cannot yet perceive.

There are ways to release the oppression of the prejudices that we collect against ourselves and others. It doesn't have to be this way. And certain questions can illuminate and open the mind to follow a better course:

- Is it possible to neutrally observe what goes on inside our own minds?
- Is it possible to respond to needs and issues from an assessment based on consideration rather than reactivity?
- Is it possible to pause, even in the heat of an argument, and listen to oneself and others impersonally?

❊ Is it possible to withhold habitual and instant prejudices about people and the situations we find ourselves in and discover more of the whole truth?

Yes, all this is humanly possible, when the will and development is there, at the point, to just hold back the tides of prejudice before they swamp our better judgement. Often this is a matter of speeds and timings, and being able to slow down enough, be cool enough inside, be present enough in the moment, to not lose ourselves in the confrontation or challenge we may be facing – to be less personal about it all.

Within this, there is an important mindset to reach for and practise, characterised in the words, 'know it, but don't think it'.

Imagine for example the circumstance in which you have to, or choose to, meet up with someone with whom you have had really bad experiences. Perhaps they deceived you or harmed you in some lasting way, or you did similarly to them. You know this, although you may still not know or have really listened to their motives, or even your own, or another side of the story, whether right or wrong. So as you go to meet them, you know as much as you can or want to, whilst clearly you do not have the whole picture.

So the crucial next thing, as you first meet face to face with that person, is to not think about what you already know. Why? Well because if you actively think about all that history, there can be no present to meet in. If you think about and instantly bring in all the hurts and wounds, there can be little allowance for healing or forgiveness. If you are so full of all that went wrong then, there'll be no space for what can go right now. At the same time, if you learn from what you know, you can know better.

So if, at the point where you are confronted by a prejudice in yourself, you 'know it, but don't think it', then you can bring yourself to a circumstance free and willing to move forward, to try again, without prejudice.

So you take your own position that carries the strength of a humanity and a reasoned attitude, rather than the weakness of your prejudices about the person and the outcome, or a dismissal of the fact that people can and do change.

You can't expect not to have or to face prejudice, but you can expect to work towards being less subject to it yourself.

Another way towards this is to try to begin each new day from the new day itself, from the now. For whatever the many tasks that today may involve, we can always determine how we want to be inside it all.

We can always determine whether we will try to be open to new possibilities, new ways of thinking, being and doing, or whether we will always fall back on the stale and entrenched habitual ways in which we may go on.

In itself, this is an ongoing personal development. It is an endeavour that begins with the sensitivity and awareness that takes us deeply into the whole premise of what life is for, what humans are for, and why to undertake a journey of personal growth in the first place. In the bigger picture of this, we may come to see that prejudice has no place in humanity's evolution, but an open mind most surely does.

There is also a particular mentality that eases the mind away from the prejudices that restrict its natural freedoms. This is another kind of sensitivity and awareness to be developed. It is one that genuinely recognises, with ongoing self-forgiveness and forbearance, that we are all born into and engraved with different prejudices. From this mindset, a greater tolerance and patience can emerge, and the ability to know what one won't do.

Consequently, this can lead us to taking a position to no longer be so governed by prejudice, releasing us from feeling the need to always be defensive or attacking upon anyone who disagrees with us... or the need to judge others because they're different to us.

We always have the option and choice to evaluate, assess and balance what we experience in life, without wielding our prejudices all the time.

Such self-leadership is a demonstration to ourselves and to everyone we meet, that we are less subject to prejudices, less subject to the shifting biases of cultural opinion and thus freer to exercise our own minds and our own free will, without prejudice.

In what ways do you think prejudice lives in you?

*

How could it be different?

Transforming conflict

If we want to wisely court and embrace the future, then our minds and practices concerning conflict will need updating.

Conflict is integral to living. It is one of the greatest personal development opportunities you can find, or that easily finds you, as is usually the case.

In facing difference and challenge in the course of our lives, conflict can be a catalyst for change, creativity and greater self-knowledge. We may suffer some psychological and emotional cuts and bruises en route, but when worked through intelligently, conflict can be transformed into new learning and a greater consciousness, collaboration and tolerance amongst people.

If not, then the destructive forces of conflict are unleashed, as they burst dramatically onto the scene with their usual 'entourage' of inner turmoil, distress and sometimes violent repercussions. Habitually, over centuries, this is what we have made conflict into.

Conflict inflicts or constructs. And we mostly have a choice in the matter – made consciously or by default. We can either deliberately use our experience in conflict to know ourselves better and progress more prudently and safely into the future, or we can end up carrying forward old wounds and scars, trapping us in our past.

The classic unfolding of conflict

Whatever the triggers, serious or trivial, conflict tends to play itself out in fairly classic, destructive ways; in which the degree of volatility and reactivity usually sets the climate and sequence in which things unravel.

The predictable and repetitive features of a conflict may seem obvious after analysis, but are rarely recognised or anticipated. Yet they need to be if we are to use conflict, rather than be controlled or defeated by it. Such classic features are:

Each person believes they are right.

Each person believes the other person is wrong.

Each person believes they have not been listened to properly.

Each person feels undervalued and/or disrespected by the other person, and/or undervalues and disrespects themselves.

Each person is unable to be neutral enough to accept that they only see part of the whole picture.

This usually leaves everyone unable to perceive a better way forward other than insisting on acceptance of their personal view, or trying to win the argument.

It is likely then that we have come to view and experience conflict as always being destructive and to be avoided at all costs. In this sense, conflict makes no sense. But maybe all the dramatic incidents of conflict that have gone on in our lives, and the way we have handled them, have obscured a purpose for conflict that we have not yet seized.

A greater purpose to conflict

The Latin root of the word 'conflict' means 'to strike together', and in the course of living it is impossible not to strike together and against all kinds of resistance, unknowns, blockages, impediments, challenges and new opportunities each day.

That's integral to how human life on earth evolves. However, whether this becomes personally constructive or not, depends on whether we are able to use such experiences to bring improvement and new ways to progress into the future.

This reveals one of the core purposes of conflict, which is **change**. Conflict can facilitate the most extraordinary changes in and around oneself.

Change can be a cause as well as a purpose of conflict. Most major periods of conflict and stress nowadays – during bereavement, divorce, separation, career moves, illness, imprisonment, moving home, etc. – are all about significant changes. We will all face at least one of these events in the course of our lives and they can present powerful opportunities to learn, grow up, change for good.

In times of change there is often a great inner conflict and tension between key influences from our past and present, and key concerns or hopes about our future. This is also true in the context of families, companies, communities and nations. And yet such inevitable conflict can be a crucial point of self-realisation, leading to changes we could never have foreseen had we not met the situations or people with whom we conflicted.

Realising that the conflict we're going through can bring about necessary changes, creates a chance to realise more about our true nature, our current strengths and weaknesses, and what we really want from life. This in turn can help us keep our choices more conscious and informed, leaving us freer to move with the future.

If we are truly seeking to evolve, we will be constantly embracing change. With such purpose in mind, we may build within and amongst ourselves the kind of awareness and intelligence that would make conflict part of our personal and collective development as humans – not part of our self-destruction as a species.

Difference

To begin to work towards change constructively and creatively, as a purpose within conflict, is to be able to understand and appreciate difference. For difference promotes change.

Yet in the middle of a conflict, the space and tolerance for difference can just evaporate. So our awareness of how we think, feel and act when there is difference can lead us to develop better sensitivity and skills in working through conflict, and in transforming it into a creative process. For when we truly recognise and respect that almost everyone thinks and reasons and sees things differently from how we do, then we can begin a process in which conflict naturally transforms into a heightened consciousness, where differences join forces in greater purpose, harmony and strength.

So if we can perceive some of the key differences that exist in and between people, and how this affects us personally, we can consciously use this great wealth of difference to develop a much more diverse and harmonic future world, unrestrained by past impediments.

Conflict arises in a very wide and permutating range of circumstances of difference, and a book could be written about each of these. But for the purposes of offering starters, we can list and dwell upon some of the key areas of difference between people as below:

Innate differences in: Gender. Genetics. Place of birth. Race. Astrology.

Situational differences in: Environment. Pressures. Balances. Timings. Speeds.

Evolving differences in: Upbringing. Traditions. Character. Role. Wants. Needs. Values. Aims. Motives. Perspectives. Expectations. Interests. Attitudes. Impediments. Stresses. Perceptions. Insecurities. Fears. Unresolved issues. Judgement. Control. Style. Comprehension. Maturity. Power. Standards of living. Communication. History. Religion. Politics. Social class. Culture and Views – especially on what is fair and just. And differences of course in people's ability to deal with conflict.

When you reflect on each of these key differences, they can all contribute quite naturally to one's personal growth, if they can be worked with wisely. Yet when we are in the middle of a blazing row in the office, at home or in public, these key differences explode into action as each person argues their point; differences in comprehension, timings and expectations for example can compound reactively with differences arising from each person's inner fears, insecurities, their history with each other and unwillingness to change – a rather potent cocktail for conflict!

Yet at the same time, if and when the row settles down, or even during it, imagine the opportunities that can emerge to discover more about your true nature, your true level of development and where change is needed. Imagine the opportunity that emerges to build new and improved relationships based on better understanding and respect.

So knowing and being aware of how differences affect us and others is vital in resolving and transforming conflict. It can bring an overarching awareness and discernment that leaves us free to choose how to respond, especially in the heat of the moment. In the middle of an argument this may lead us for example to simply say, 'It's clear that our expectations, values and understandings are very different here. How can we find a better way forward?'

Allowing space for difference, and developing tolerance in the midst of it, is a matter of training. And it can begin by knowing oneself better in the face of difference. Upon this foundation a higher elevation becomes possible, where conflict transforms into a creative future-bound process of change.

Mindsets and agreements that can transform conflict

In the shocks, upsides and pitfalls that are often part of conflicts between people, one person's account rarely sits comfortably alongside another's. In which the third side to the story remains untold when neither party manages to see, or wants to see, the bigger picture and context, and when the sheer stress of a conflict masks the real issues.

In some instances the conflict going on between two people may not even be directly related to them personally; it may be what they have become subject to through an ongoing family, neighbours' or workplace dispute, or even a war.

Whatever the case, behind the who did or said what, when, how and what was meant, intended or not, there are certain fundamental, practical mindsets that can assist in transforming the whole way we approach, perceive and deal with conflict in ourselves and with others. Such mindsets as those which follow call for contemplation, and making real in oneself:

- Every human wants to express themselves and be listened to with respect, and without judgement, censure or blame.

- One can only be accountable for oneself. And whether others want to resolve conflict or not, one can always work through and resolve conflict in oneself. This is a matter of choice and personal development, which can lead to conflict being transformed into self-realisation.

- To transform conflict is to have developed a working attitude in oneself that understands its place in human development. It is to work with difference, towards changes that are productive. It is to proceed without unrealistic expectations that everything will quickly or perfectly get resolved, but with the expectation upon oneself to be constructive, patient, fair and reasonable, irrespective of how this is met by others.

- It's vital how you talk to yourself about the conflict you're in, and how you reason what is really causing it and what you are learning from the conflict – especially about yourself. In which knowing what you will not do, and the triggers, reactivities and allergies that live in you about people and situations, can create greater safety and skills in yourself and within any dialogues. Towards this, you may want to get some accurate feedback about yourself from someone neutral and knowledgeable enough.

- Realising the bigger picture at play puts conflict into context, brings values and other priorities to the fore, and helps render unimportant matters unimportant. Within this, it's really constructive for each person to take some time to refresh themselves about what they value as being most important about life, and about their lives, and about people – including, if possible, about what one values about the other people involved in the conflict. It needs working out.

- If conflict is an opportunity for anything, it is certainly an opportunity to consciously develop human qualities in oneself; qualities such as patience, restraint, openness, consideration, forgiveness, tolerance and understanding. The quality of consideration for example, can engender a more reflective and compassionate process into dialogues, where reasoning and listening prevail.

- Trying to change someone's view, or enforcing your own, usually prevents agreement or forces one that won't last. There needs to be recognition, space and respect for each other's differences, as well as a consideration of the different realities that each person is facing.

- There also needs to be the conscious awareness and understanding that each person will likely be feeling all kinds of emotions, hurts, unsettlements, fears, sadnesses, frustrations, anger and so forth; other friends and family members who are aware of the conflict may be feeling the same way too. There's no point in suppressing such anxieties, but there is a point in not allowing them to fester, and in letting go of what becomes toxic. Within this it's important to not be overruled by your feelings, or to try to recruit everyone to your side of the story, for this curtails or nullifies reasoning and resolution and can lead to escalation and retaliation instead.

- If you believe an apology from yourself is fair and needed, then why not offer one, without expecting one? It can make you clear with yourself and bring easement all round. Responding freely to a real need is a great strength.

- Taking oneself less seriously can work wonders, as can natural moments of humour. Many conflicts, when you detach yourself from the storm, can actually be quite funny as well as educational.

- Bringing the best of yourself to the situation, and upholding the best of others, may be an honourable stance you want to take – irrespective of the outcome or how anyone else is. This calls for exercising self-discipline and being open to a different view than the one you come with. Ultimately, it's a matter of what kind of person you choose to be.

With such mindsets and qualities at play, each person may be better able to find in themselves, and together with the other person, the kind of **agreements** that could facilitate a resolution or transformation of the conflict, such as agreeing:

- To create some clear and calm space and time for each person to freely express themselves, and the issues they see to be importantly at play.

- To share why and how one is seeking to resolve or transform the conflict, and what one's hopes and aims are in this.

- To put in place some basic standards of engagement, such as monitoring the process so that it doesn't speed up or heat up beyond breaking point, and taking time out for a cooling off period if necessary.

- To work through the issues without blame, constant interruption or expecting it to be perfect – yet with a willingness to listen and speak with respect.

- To meet in a place and at a time best suited to everyone.

- To call upon a mediator, or someone trusted by all parties, if the need arises.

Although not comprehensive, such mindsets and agreements are truly pivotal. They enable us to resolve and transform conflict into a constructive, learning process. They enable us to stop, move into the long view, and thereby improve the ways in which we respond to conflict as part of our own development as men and women. In such ways we may contribute to humanity.

Working through conflicts successfully is a versatile, patient, reflective and illuminating process, albeit often a turbulent and exacting one too; but it's not nearly as exacting upon us as holding on to painful memories, regrets, unclarities and emotional wounds.

In the end, we either let conflict make us its victim, or we learn to use its resistance to open and transform our lives with new insight; ultimately, to effect change and move on, from conflict to confluence, from opposition to proposition.

For conflict can help bring us inside the changes that we need, inside what and who we really are now, and inside what we truly want in life. Therein lies a very hopeful future, where we use the resistances that we meet in life to grow our essentially spiritual and compassionate nature as human beings.

How can you deal with yourself better
in times of conflict?

*

What new ways could you now develop
which might transform your conflicts with other people?

Understanding and dealing with cynicism

Human experience forms an amazing mosaic, with smooth matching pieces, missing pieces and ones that get so worn down that they don't fit any more. And those jaded and faded pieces can make a very distorted picture of life.

For our experiences in life may be viewed as good or bad, but it's how we consciously understand and use them that is either constructive or damaging to us. Of course if we take all of our experiences too personally, we'll likely end up somewhat cynical.

So do you consider yourself to be cynical? Perhaps the present-day culture, with all its hyped-up hopes and expectations, and inevitable disappointments and depressions, leaves us all a bit cynical. But what really is cynicism?

Well, the Cynics of Greece from around 2,500 years ago seem to have acquired a poor reputation in our times. But these 'founding' Cynics, such as Antisthenes and Diogenes, simply rejected the pleasures, profits and materialism of their time. They embraced the pursuit of virtue, which had little room for possessions – other than perhaps living in a wooden tub on the streets of Athens! So perhaps originally, being a cynic was actually something virtuous to aspire to.

Later on however, some of those who followed the Cynic way of life appear to have occasionally used their indifference to society as an excuse to borrow money and be somewhat unconcerned about repaying it! No wonder the respected view about the Cynics turned, well, a little cynical.

Cynicism for us in these times is mostly defined as a scornful negativity, distrust or jaded view about something or somebody. But to really understand it calls for a more neutral approach.

It's easy, and perhaps even warranted, to be cynical about a leader or politician who is promising the earth, whilst one feels and reasons that what they're actually promising will cost the earth.

More importantly, as one experiences more of the choppy, turbulent cross-currents of life, and the setbacks and self-doubts, one can develop a certain inner bitterness, recalcitrance and a lack of belief that things will ever change, improve or be fulfilling. And it's easy to see how a person would arrive at this, given the fact that in the electric, pumped up, consumerist societies of today, we are sold a dream of success that is usually hollow when we attain it, and full of despair when we don't. Thus the never-ending pursuit of acquiring quantities of things, friends, congratulations.

Personal relationships seem to breed more cynicism than anything else. How quickly do great expectations rise in a relationship and then collapse in a whirlwind of bitterness, misunderstandings and disappointments? Cynicism blinds us to a very simple and crucial truth:

The deeper, lasting feelings of success, happiness, fulfilment and achievement are to be found at first, at core, inside oneself, and not through another person or job or thing.

You may derive your self-view and self-worth from what society and everyone else reflects back to you, or from what you reason is the case, first to yourself. And it's important to do such reasoning. For someone can say to you, 'You always seem so happy' or 'You are so compassionate' or 'You are so shallow', but only you can come to know and live the truth of this and discover whether or not this *is* the truth… the truth always being more than people will or are able to reflect back to you.

This reveals another important insight, *that cynicism is an unnerving disappointment, which festers and hurts in the gap between our unrealistic expectations and aims, and what we then experience and discover to be the case.*

We may become cynical about governments because we keep over-expecting and they keep under-delivering. We may become cynical about achieving what we expect, aim and hope for in life, because our experience reveals that maybe we can't get it, or that if we do, it won't last or be fulfilling.

And when we become too cynical we get lazy, because it begins to sap our energy and will to live and causes us to be too personal and opinionated... which in turn can make us even more cynical. So maybe we can learn something from the early Cynics by asking ourselves a key question: *do we really assess our self-worth by our job, our home, our friends, our opinions?*

Or do we firstly value ourselves because we value the precious gift of the life we've been granted, because we value being human and being able to exercise choice, and because we feel a profound gratitude for life itself which transcends the daily stresses and superficialities of modern society?

Cynicism simply depletes us, as it delivers the tired seen-it-all-done-it-all-nothing-will-ever-change stuck record, which dulls and taints natural feelings. And it only takes a few setbacks in relationships or at work, the tragic death of someone close, a feeling of being unappreciated and undervalued – so many different triggers – to create an atmosphere of disillusionment or despair inside oneself, with oneself. But we can determine a change of outlook, if we choose to.

Through all our mistakes, disasters and perceived failures, nothing can destroy that quiet space, deep inside us all, where we still believe, we still love, we can still feel passion and excitement and the mystery of what comes next in the unfinished story of our lives on earth. This is sacred space.

So whatever the depressions of yesterday, we can always take heart today that there is that knowing, steadying, still place to be found inside.

And when we take the time and care to reach that place, we may look at life with the freshness of now, no longer constrained by the seemingly overwhelming burden of what may apparently be wrong with us.

Setbacks will always be a part of growing wiser – if we can learn from them. In which the conscious agreement with oneself to learn from mistakes makes us less punishing about making them.

In daily living, it becomes important to be able to take some time out to look at oneself again, afresh, taking care to see the qualities that we as a person have actually built; a time to recall and refresh the respect we have for our lives, for life itself. In such a way we can get our heads above the quicksand of a world where cynicism is often used as a sign of mature adulthood and being worldly-wise. It truly isn't.

Cynicism is a surrender to a predictably jaded view of life, reinforced by the worst of our personal experience.

But we are the authors of our own destinies, when we no longer fear what others think of us and become free from peer group habits and from lifestyles ruled by dogma, apathy and familiarity. Being released from the bindweed of cynicism calls only for the liberation that is engendered by such reasoning and perception.

For in this light, cynicism reveals its bitter and casual leeching of the inherent will and passion to live life to the full. It doesn't have to be that way.

What does cynicism feel like?

*

What does not being cynical feel like?

Being intact is a freedom to let live, and let go

Intact

This is about
... the intactness of the universe, humans, fish, cells, genomes, atoms and all kinds of living things that can be known about without slicing them to bits.

This takes the position that
... in spite of the pressures and persuasions of these times, which can addict us to an inflated sense of self-importance, we may still attain a freedom to be what and who we essentially are as human beings. Therein lies the search for meaning and the fulfilment of being useful during our brief stay on earth.

This speaks of
... an unforced, open, willing, responsive way of life, instilling trust, self-respect and a sense of completeness. This can connect us to the qualities and ways of the very Creation we are born into. It can lead us to become co-creators, partners with life; in it for what we can contribute, not just for what we can personally get out of it.

In such ways we may become and feel whole... in touch, inspired, intact.

Being able to consciously let go of some of the debris of living – the hurts, guilts and anxieties – is an essential art in staying intact, balanced and more at ease inside. Not that one doesn't still process regrets or anger or shame, but the aim is to be able to let go enough of the experiences that carry these wounds, to reach a point at which they no longer dominate or control our lives.

This calls for a process of good reasoning, to quieten, relegate and settle inside what needs settling. Such reasoning involves being honest, fair, logical and not hard on yourself or others. In which some quiet moments of meditation each day, to affirm what we really want in life and what we don't, can help greatly towards this state.

Fundamental to this is an ongoing self-forgiveness and release, understanding that the world we are growing up in, and from which we accumulate so much debris, often leaves us engraved with so many troubling insecurities. And this particular lack of intactness breeds an ingrained and very reactive, defensive and/or attacking mentality – a kind of 'indignation at the ready'.

Such a mentality is printed in us at a very early age, because of the bombardment of assailing and pathogenic impressions – from noise, conflict, daily tragedies in the world, to the judgements of others upon us. Within this is the onslaught of aggressive competition, pressure to be the best, most perfect, popular and attractive, which can cause us to develop a constant attack mentality as we press hard to get what we want, or what we think we want. Conversely, this may also cause us to feel that we have to defend and prove ourselves all the time, just to survive it all or to feel some self-worth.

The vacillation between defence and attack is an extremely debilitating, fragmenting and stressful ongoing process. It can leave us brittle and unforgiving, making it almost impossible to stay intact.

This is why good reasoning and self-forgiveness are such important steps in the letting go of this ongoing turmoil and in the achieving of a healthy intactness. And part of this is the release and relief of taking oneself a little less seriously, a little less preciously, and throwing off the shackles of our unnatural, unhealthy, over-blown expectations to be regarded as more than we actually are.

This can lead to an inner settlement that involves the application into daily living of such reasoning and ongoing ways as these, elevating beyond the polarisations of attack and defence.

This in turn can develop a peace of mind that comes with being fair, honest and reasoned with yourself, bringing a growing pride in actually being human, and loving what a human can be and do and become.

It's one thing to read such words, or even agree with them, but quite another to stop, relax, reflect and feel the release of your own mind as you give it permission to really appreciate the fact of being alive, now, inside one of these amazing, hugely unexplored, powerful, artful, beautiful human forms.

A science of intactness

There is a science to this, for the inner settlement, healthy balances and well-being that make up intactness don't happen just because we'd like them to. Intactness is a natural resultant of applying care and patience, whilst seeking to understand the truth of what we meet in life. It comes from settling to and knowing what you can change and what you can't.

Developing a greater intactness is therefore a natural, human life science that leads to a greater knowing of self.

Science is a faculty inside the universe or 'university' of a human. As such it is an innate capability to discover and know truth, through search, research, study, experience and living. For many reasons however, some scientific approaches have been about destroying the intactness of living things, either to make commercial products, weapons or simply for acclaim. Some of the largest international scientific projects of our times – whether into space, genes or atoms – have claimed some successes, but at what cost?

If we look at the origination of the word 'atom', which comes from the Greek meaning 'not divisible', we might ask ourselves whether an atom was created to be split from its intactness.

For what kind of a future lies in the science that would research and split that atom (nuclear fission) in order to create increasingly powerful nuclear weapons that can destroy increasingly large numbers of people in one go? And do we really need to smash particles inside miles of expensive tunnelling to discover a so-called 'God particle' out there, when there may already be so many to discover in here, in us, freely?

This does not take a position against mechanised scientific experimentation which is clearly important when it reveals something new and useful without doing any harm.

This takes the position that studying the macro worlds in the micro, and vice versa, does not have to begin with the taking apart or destruction of anything, because the observation of living things in their intact state can reveal much more.

In which it's worth reflecting on how it was that Greek and Indian philosophers already knew about the atom over two thousand years ago, using the technology of their minds.

When you dissect the intactness of living things, or remove them from their natural environment, you can lose the ability to understand their significance, function and greater context. And you may end up devaluing life, which only holds its value when it is intact, whole.

It's like the difference between studying the behaviour of animals in the wild, or in cages in a zoo.

Do we therefore need to question the mindset and thinking that can now deliver to our dinner table, as one example, a genetically engineered salmon, rapidly matured, several times larger than normal, and with a gene from a different variety inserted purely for profit? Are the consequences of such actions in the bigger picture being understood and prioritised? Or is the wisdom of long-term thinking being blocked by the perceived need to deliver a result now, whatever the cost to future generations?

Is it a fundamental lack of intactness within us as humans, and a lack of the settlement and pride in being human, that causes us, for example, to try to genetically engineer or modify life to somehow improve on it?

We cannot create life, so why would we not respect and revere whatever it is that can?

Nothing that has been naturally created can ever be fully understood by interfering with or destroying its natural living template and function – for then we no longer see it as it actually is, in its intact state.

Today there is so much knowledge and so much writing about how all living things on this planet, in the skies, in the galaxies and in the universe, are absolutely interconnected – from one atom in one cell inside your bloodstream to the substance of far distant stars.

We may know this, but perhaps it is again a lack of inner intactness, individual and collective, which has so disconnected and destabilised the intactness of most ecosystems on this planet. For a lack of intactness can leave us feeling fragmented, out of sorts and in conflict inside, and therefore disconnected or antagonistic to greater needs in the world outside ourselves.

Being intact

When we are intact or in touch with our true nature, at ease, well within our own being, and acting consciously from that state, our approach to life can become much more tolerant, considerate and selfless. This in itself can contribute so much to the greater intactness of a family, a community, a nation, humanity.

Being intact is an ongoing personal development, a daily quest to discover, improve and be guided by what the natural worlds and natural laws teach us every moment, about harmony, balance, order, constancy and co-existence. To be intact is to consciously observe this, to genuinely appreciate it and to become part of its natural well-being and purposes.

It is out from such an endeavour that a person can really feel whole and well and settled with one's unique chance at life; for this builds a still, quiet core inside oneself that rides with the setbacks and wounds and fears, whilst remaining intact.

What does being human really feel like?

*

What might allow you to become more intact
in yourself?

Climate change in the human

We are born philosophers

Yes, the word 'philosophy' means 'a love of wisdom'. And we show our love of wisdom at many levels... whether it's about the best way to develop our gardens, our relationships or our intuition. Ask anyone in the right way, adult or child, and you may well discover some great wisdoms about human psychology. We all attend the university of life. We are born philosophers, practising or not!

Philosophy is not, as a first principle, an academic subject. So do we need to begin with who Socrates was – and what we think he thought but never wrote down – to start to understand philosophy? Do we have to know what Thales thought was the substance from which everything was derived, as opposed to what Democritus expounded?

And does it really matter whether philosophy is considered to have started in Greece 2,500 years ago, or in China or in Ancient Egypt or wherever? Great people and great civilisations are not necessarily a great place to begin.

Could we start then with the fact that as philosophy is a love of wisdom, it is innate in us all?

Then we can begin inside the reality of our own lives and living, as opposed to immediately leaping to the assumption that philosophy is only 'out there' in some essential-to-read book, major theory or famous saying. In the quest to understand life, these may offer profound guidance, but there is a more essential starting place.

Philosophy begins in the instinctive urge, quest and desire that stirs in us all from a very young age to question life – to question everything! It is a human endeavour to discover life and understand why things are as they are, what they do, how they work and how to live our lives to best effect.

Philosophy is an intuitive striving to be wiser, more perceptive and more effective in the way we conduct our lives. It's a seeking to deepen, enrich and develop the reasons and feelings and actions that we make and that make us who and what we are. We each have a philosophy about life – whether conscious or not – to be discovered when you stand back and carefully study the way you actually lead your life… it's right there.

You may, as an example, have a philosophy rooted in the importance of self, where you consider that 'being the best' or being highly competitive or being liked and confirmed is a satisfying way to go on. Or maybe you have a philosophy in which you call yourself an optimist or a pessimist or a realist or a believer, non-believer, just-want-to-be-happy-and-make-the-most-of-things.

But maybe it runs a little deeper in you. Perhaps you have developed a philosophy rooted in a deep value for others, in which you aim to be honest, generous and respectful. Or perhaps you have found a deeply searching philosophy, where you freely explore the true purpose of living, seeking to fulfil this beyond just surviving in a materialistic society. This may cause you to want to open up and develop the range of human talents and ingenuity, to understand some of the great mysteries in life, and why they are mysteries…

We are born philosophers, because we are trial and error creatures who have an inborn love of doing things with increasing wisdom rather than diminishing wisdom.

And for some people, the honing and refreshing of the ways and principles by which they lead their lives, is a serious and inspiring undertaking – a philosophy of self-realisation and liberation.

For such philosophers, their greatest tools are:

❊ *An awe and reverence for life and an ever-searching outlook.* This cannot be learned because it is already in us. But it can be consciously freed up and developed, especially by constantly looking for what is significant in life, disregarding what doesn't really matter.

❊ *A humility* to know that we mostly don't know (thank you Socrates!) A humility that understands that we know so little about the universe out there and within ourselves, and that the complex human world of imaginings, feelings, prescience, attitude and psychologies remains largely unknown to us – but absolutely invites exploration!

❊ *Questions, questions, questions.* These are the most amazing and effective tools a human can apply. Finding the right questions is a great art, in which the six question 'power tools' of why? what? who? when? where? and how? are perfect for the job of living – if they centre around the reason why. And don't children already know this instinctively?

❊ *A deep desire to discover, explore and research* the great ideas, the great issues, the great meanings and arts in being human.

❊ *A love of the unknown.* What's the point in fixing ourselves within the confines of what we already know and can already do? Where's the progress in that? Why would something create a human with the most incredible range of faculties, only for the purpose of living the most inhibited life?! That would be like owning a huge fifty-room mansion but only living in one of the downstairs cupboards!

Philosophy is not a dry-biscuit, opinionated book study, or a theoretical, hypothetical, intellectual exercise. It is a natural inborn passion to explore life wisely and bravely, with a determination to live by the very wisdoms that you discover and that you say you love.

So, what is your philosophy about life?

Three faces of the human

We seem to have a fascination with diamonds. Maybe it's because we see some of ourselves and our finer aspirations in their outstanding beauty. And the natural development of diamonds and humans share a similar growth pattern...

We are born with a latent conscious intelligence that we apply as we make our journey through the many different impressions and pressures in life. This shapes and sculpts our character and demeanour. So we begin life very crudely, just like the carbon first stage of a diamond – indistinct with distinct potential.

Then early experience through childhood, with its urges, pressures and intensities to explore life, may start to hone a small 'rough diamond' inside us. In time – through training, self-discovery, illumination – we may naturally come to see our lives in a different, rarer light, and become finer through it. This coruscating light reveals the higher possibility of human brilliance, radiance and brightness, in its many different facets, reflecting what and who we are.

The mystery and attraction of diamonds revolves around their ability to absorb, process and reflect light. Indeed diamonds are evaluated by their clarity, their luminescence, their fluorescence and their lustre. This capacity to radiate and transmit light is diminished by flaws and inclusions. And so again with us humans, for like the diamond we can shine and have the clarity and lustre of a fine radiation, character and presence, whilst impaired by the flaws we each carry.

When we are lit up, we reflect the light of life more brightly, in which our decisions and the positions we take are like 'cuts' upon our 'diamond' selves. And the cut of a diamond is always a determiner of its brilliance.

Therefore what fashions the faces, phases and facets of our lives is in the nature of what influences and forces we allow to use us in our motives, feelings and actions. The more these are flawed, tainted or inhibited, the less bright we may become. The more these are true, clear and potent, the greater the illumination.

So we become the face of what lives in us, what moves us and what lights us up, or not.

And amongst the many subtle and not so subtle powers that flicker across or radiate from the inside to the outside of the human face, there are three distinct elevations that go on in us – each inter-relating and interfacing. You may observe these three faces through and on yourself or another, and whilst the mirror reflects this first one below, it only offers clues about the two that follow.

The default face

('Default' like the basic settings on a computer)

The default face is the character of physical face that each person holds when 'at rest' or when no one is watching. It is the face of what is and isn't moving through a person when they are not self-conscious or animated in conversation, or simply when they are on their own.

It is the face that shows our fixed expressions or 'settings', which changes slowly over time and is engraved with a whole life's experience as well as what is happening in us in the moment. And mostly we are not aware of this face, as we don't see ourselves physically all the time.

A particular nationality, religion and/or cultural group can be seen to possess a common default face, because of the similarity of history and influences that tend to affect each person in that grouping. So the default face will have features or perhaps strong traits that come from the core traditions, character, genetic and heredic trace of that grouping.

Over and above this, the default face can primarily be one of hope or disillusionment, willingness or anxiety, determination or subservience – mostly a particular blend of many qualities that are all residual from what we make of our unique experiences in the world. And very often in our times, the default face on a person has a notable blend of strain, worry and longing as its constant featuring.

Skin-deep beneath this face often lurks a layer of defence or attack, for the gripping insecurities of the times we live in, and the stresses of surviving in a modern world, tend to cause us to depend upon attack or defence just to keep body and soul together. So the default face might frequently carry the tensions of insecurity, or a lack of peace of mind, or the upfront pressure to prove oneself and be confirmed on a constant basis. Alternatively, it may reflect a person's optimistic and carefree approach to life.

The character of what the default face is presenting is there to be detected, especially in the tensions across the skin and its overall tone.

Whatever the fundamental nature of the default face in a person, it is curious and quite extraordinary that at death this face often resolves into a peaceful one. For once the accumulated inner stresses and pains are released and the taut tensions ease, a joyous or even sublime look may appear. Of course, one doesn't have to wait to die to show this face!

So the default face is visible in the lines, features, texture, tone, shaping and contours of the physical face.

And, as we read the character of this face, we read some of the end results of how living experience has formed and influenced that person in their life so far. Yet it's important to realise that not everything on the inside is portrayed on the outside face.

The face of endeavour

This face is behind the face we see in the mirror – although the physical face displays some of its nature. It is the face of what is inspiring or depressing, energising or depleting a person. It itself is unseen, yet it appears through the default face of everyone, however inscrutable they may appear.

What permeates through the face of endeavour are the inner qualities, urgings and longings that we possess as a human. Yet like an imperfect image on a screen, this face also displays the interferences and distortions that we allow or cannot avoid.

You sense this face, especially when you look deeply into a person's eyes – beyond their appearance, body language or facial expressions.

It's perhaps this face that shines through the eyes and smile and look of the Mona Lisa, which captivates so many people. To achieve the depth that brings forth this subtle 'inner face', Leonardo da Vinci used layer upon layer of paint with great finesse.

The face of endeavour can feel inviting or repelling, hard or soft, deep or superficial – often a blend of all these and more. And this face can change rapidly depending on the change that may take place in a person's endeavour in life.

For example, someone suddenly finds a great passion or thrill in their life, or a new interest or purpose of some kind and develops an outward appearance of happiness or well-being, which overlays their grumpy or dejected looking default face. For an inspired outlook and attitude about life radiates through the default face, and is attractive and charming beyond physical looks. By contrast, if a person's work or relationships are failing badly, their disappointments may also show through, even when they're smiling.

So this face represents what a person is endeavouring to be and do with their life. It expresses their search and discovery and aims in living, whilst showing the struggle and challenge and resistances being faced or overcome. It carries the content and 'feeling life' of their experience.

We tend to judge people quite instantly by their default face. But it is when we perceive the face of endeavour behind it, that we begin to know more of what is really going on inside a person's world, and what they are striving for.

The face of illumination

This face is more subtle again. It is the face of what causes us to do and to be the way we are. It can also be the face of something greater than us that we are connected to and that is moving through us.

This face reflects the motives and reasons of why we do what we do in life, and what cause or causes these motives are connected to. And the level, power and value of what this constitutes, determines the degree of illumination, or not.

If what motivates a person is not constructive or generative or bright about things, or has become dulled and disillusioned by experience, then little or no illumination can sustain. There's no morality in this, it's simply how it works.

On the other hand, strong, compassionate and clear motives and values in a person – lights such as these – can make for a fine illumined face that is much more an emanatory radiation than a physical portrayal.

You might have seen such illumination through the face of someone like the Dalai Lama or Nelson Mandela, or a friend, a stranger, or yourself. It carries a fine radiance, the stature of being with something greater than oneself, and a long view in the eye as can be seen through the Egyptian eye, or eye of Horus.

A naturally illumined face that shines through a person does not stare, confront or pry. It is clear, clean and not demanding, judging or seeking satisfaction of self as its first principle. For within its shine is a well-being and self-determination to live and respond to needs in the moment, with a vision of what is yet to come.

When this face within us is brightly illumined – like sunlight streaming through those huge cathedral rose windows – it radiates an ethereal quality, a radiant image from the source and cause of life. At this level it will not put its face upon the face of what it meets.

So many painters around the world have tried to depict this face as an iridescent radiation, a bright halo, a field of energy or a shimmering, colourful aura around a person.

However, like the very finest and most valuable of diamonds, the essence of this face is naturally colourless.

It is these three faces, all active, separate, yet permutating together, that reflect three core aspects in human existence. These can be characterised as cause, process and result:

Cause: This relates to the *face of illumination* which radiates the causes and motives that we stand for and represent, in the ways we lead our lives.

Process: The *face of endeavour* reflects what we process in our striving and application in living our lives, and the qualities and values that we embody.

Result: The *default face* represents the end results of all this, which become etched onto our outward physical face and the features it displays.

What is on the outside face of a person reflects what is on the inside, but what is on the inside is not always apparent outside. So the ability to see and understand these three faces depends on whether we have trained our eyes to only fix on end results, or whether we have the care and vision to see through this to the causes and processes that go on at much deeper levels in people.

A person who is two-faced, uses the default face to hide a deceptive ulterior motive, lurking in the face behind – in which the face of illumination simply goes dark.

In a person of integrity however, the three faces light up as one.

Can you see these different faces,
on and in yourself and others?

*

How can this help in the better understanding
of yourself and other people?

Humility
and the admission of fallibility

Today's societies attribute great virtue to knowing a lot of facts, however liberating, limiting or chaotic this may turn out to be inside our heads. And around the world there are many popular quiz programmes to make us feel good about this, especially when we get the answers right and get confirmed by how many facts we know.

But we humans can become obsessed by the desire to acquire and deal in facts alone, often very trivial ones, to the point where the truth behind the facts becomes ancillary.

The following simple fact however may be helpful in approaching humility:

Every human being alive on this planet knows a very, very little something about the inconceivably big everything that there is to know.

And a truth behind this is that whatever caused you and me to be created as human beings, clearly meant for us to be a vastly unexplored, infinite blend of possibilities and opportunities, way beyond anything we can envisage.

For the universe that is the human, is a tiny microcopy of the great universe we are part of. We humans, this galaxy, the billions of galaxies, endless space, time, other dimensions and the whole teeming mass of universal life, remain virtually unknown and unimagined in our minds – and yet we are still a living part of it all.

If this is the truth and knowing and awe that lives in our minds and appreciation, as a close awareness and realisation, then perhaps a fine, sensitive web of humility quite naturally begins to weave its quality directly into our character and sense of who we are.

With this web of humility growing inside our considerations and outlook on life, as a deliberate developing consciousness, we may become freer and more engaged with what we don't know, than in what we already do. This then grants open entry into whole realms of new understanding and progress, and into the unique proposal that our individual lives in the universe make.

Such a natural development, elected into, may cause our lives to exist right on the leading edge of life itself, where the unknown is seeded and blossoms into the creation of new and original life.

Humility may therefore feature a certain humbleness in and about oneself, but this is not the essence of it.

If humility is rooted in any one state in the human, then perhaps it is the state of acquiescence. This is an ongoing way of life that is deeply accepting, reverent and receptive to the natural ways and purposes of all living things, rendering us neither familiar nor assumptive with the precious moments we have to live.

When such an active humility inside us meets the next experience, the next person, the next conversation, its first approach is truly open – meaning that we will tend to listen to and assess what we encounter with greater questioning and care. In which the blue touchpaper to the illuminating fireworks of perception arising out of humility, is ignited by that beautiful self-statement and liberating three words, *I don't know.*

In further contemplating this deeply centering and freeing human quality, the following cascade of reflections overflowed…

❊ Humility is an ongoing admission of fallibility, and an ease born of not trying to be perfect.

❊ Humility is like the gentle warm-up that you know you need to do before you can really run, for this is your mind giving easement to your body, in the humility of not assuming, nor oppressing your systems into immediate action.

❊ Humility is like the wide-open eyes of a child, playing with a branch, a bucket of sand, a hat; joyfully sensing, wondering why, how, wherefrom…

❊ Humility causes a person, when in conversation, to listen intently, deeply, without pre-empting with the next thing they want to say.

❊ Humility is a care to not interfere with or spoil the true order and way of how nature shows us it wants to be treated.

❊ Humility is a sacred acquiescence, a welcoming feeling, and a profound reverence. It grows from the throughout appreciation that one's life has been granted a supreme freedom to be and do, gifted as it is with the powers of consciousness and conscience.

❊ Humility is not weak, not a subservience, nor a tentativeness. Humility is a strong resolve to be fair with oneself and others.

❊ Humility is in happily, seriously knowing that you will make mistakes – and in wanting to learn from them. It is an appreciation of the full value of being a trial and error functioning human, who can improve in the next moment.

❊ Humility is a tasking to be human, to be proud of it, and to not inflate that pride above its station.

❊ Humility is in the quietness and peace of wanting nothing, except for life to be as it was created and meant to be, and finding ways to join in its purposes.

❊ Humility flourishes when overcoming the feeling of having to prove oneself all the time.

How does humility live in you?

Whatever possessed you to do that?

'Whatever possessed you to do that?' she cried.

He looked puzzled. 'I just don't know. Something came over me. Something took me over, and when I came to my senses there I was wildly thrashing out at this guy. All he'd done was cut me up in traffic.'

How many times, in small or major ways, does something very powerful just take us over? Good or bad. A violent burst of anger can cause us to lose our mind, and yet the overwhelming passion of suddenly falling in love can cause the same.

What is that something that can possess us to do the things we do?

Well, if you stand back and look at one of the important ways in which human beings actually work, you discover that we are influenced by different pressures. Our language signposts these pressures in the terms: impression, expression, depression, compression.

We receive impressions and make expressions – a flow of taking in and giving out. And when the living flow of this is prevented, it can lead to depression as in feeling low, or compression as in feeling flat.

So we go through changing states of high pressure and low pressure, as does the changing weather. And like the weather, when either the perfect storm or the perfect calm gathers in our life, our habitual patterns of behaviour tend to disperse. We may then enter a world of extreme weather in our behaviour – constructive or not – in which we are caught unawares.

Later, when we reflect back, we might shake our heads and say to ourselves, 'I just can't understand why I would ever do such a thing. Whatever possessed me to do that?' Well a first clue lies in these variable pressures and influences and the overall climate inside ourselves, which can radically change in an instant.

For whether we are aware of it or not, there are always influences that occupy or pre-occupy us, for good or ill. A very competitive environment can influence us into extremely aggressive behaviours in the same way that working amongst people who care can influence us to be more tolerant.

This is even more so when groups of people get powerfully influenced beyond their own individual constraints. Think of the wave of hope and optimism of the sixties that swept through and possessed a whole generation, or the waves of despair that preceded and punctuated events in World War II.

Much more subtle states possess us too, such as hesitancy, longing, compassion, suspicion, disquiet.

Possession, as it relates to the human condition, simply describes the controlling state within a person. It is what we allow ourselves to be governed by.

This doesn't have to be either something out of a horror movie, or a Shangri-La fairytale. For a person can possess a particular quality, skill or aptitude as much as they can possess a foul temper, bad attitude or fanciful imagination. Equally, a person can find themselves unwittingly taken over by influences outside of themselves, such as waves of anger in a demonstration, hysteria at a concert or jealousy in the workplace.

The key question is, *how self-possessed in life are we?*

In other words, to what extent are we conscious, directive and aware of what happens in and through our lives, and how much are we creatures of circumstance, events and the vagaries and persuasions of the societies we live in?

There is a difference between watching, balancing and surfing on the natural currents and waves that each new day brings, or just bobbing up and down like a buoy.

And beyond, lies an extraordinary, higher world of possession. A world where finer influences, forces, knowings and feelings converge within a person to create a wholly original state of affairs that can connect to new intelligence, new ideas and unprecedented abilities.

Think of the convergence of the set of timings, influences, needs and reasons that caused the human to discover the wheel, conceive of and build a pyramid or an aeroplane, compose the music of the spheres, dance the sacred dances, paint the most beautiful paintings or achieve true inner stillness. Such is natural genius – something not only possessed by the few.

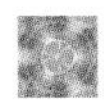

The deeper meaning of being self-possessed, as in maintaining one's mind or soul in a fine state of equilibrium, is at the root of all high, spiritual, personal development; for such states are illumined by great presence, and the essences that are naturally meant to live with human beings – such as the essences of love, honour or willingness.

Ultimately, the human is designed to possess and be possessed. Not by things, or influences that harm – for these are unnatural to the human design – but by such qualities as forbearance, constancy, curiosity and an enhancing range of creative forces and feelings.

Which then resolves into a matter of choice, and leaves us with a deep contemplation:

Do we always exercise choice of our own conscious free will? Or are we often conditioned or persuaded in such a way as to make us believe it was our choice?

So, whatever possessed you to do the last thing
you were really not pleased about...

*

... and the last thing you were really proud of?

Everything we do to ourselves
and each other
we do to the planet

Climate change in the human

Climate change is often cited as the single most important issue facing humanity. This is the message from climate experts, leaders and organisations the world over. But what does it mean? For in the next breath we seem to be straight into the cut-and-thrust language of carbon emissions/footprints/trading, 'green' taxes, solar panels, wind farms and so-called ethical projects by major energy companies.

How did we get to a situation where after over two million years of human evolution on this planet, we now have a window of perhaps only a few years to limit the damage humans have done to this earth and thereby to themselves? And can the damage ever be reversed now?

Why is the human race destroying this planet? Could it have something to do with the way we treat ourselves and treat each other? The equation is simple: when humanity becomes increasingly inhuman, destructive and divorced from its true nature and purpose, we increasingly destroy the natural worlds around us.

What kinds of pollution for example are spread by minds that promote weaponry and wars and their devastation of whole neighbourhoods, families, cultures, forests, animals? And what of the soulful cries that go out every second of every day, as millions of children, women and men the world over try to live through another day of fear or torture or the desperate lack of water, food, shelter, peace?

This climate must change first.

First principle

With climate change, many immediately assume that we are talking about the planet as a first principle. And so the critical changes that are needed in human behaviour, attitude and values often become an afterthought.

Although climatologists now mostly agree that human activities, rather than natural variations, majorly drive global warming, these activities are defined as the effects of over-industrialisation, the burning of fossil fuels, deforestation and the paucity of galvanised human response to this across the globe. But these are alarming symptoms – as critical as they undoubtedly are – of a much deeper issue.

Since the Industrial Revolution, which began in the late eighteenth century, there has been a radical change of climate in our lives and our lifestyles, with an explosion in the manufacture of products and services to supply everything we now expect and want – whether we need it or not. Overall, people in heavily industrialised nations have more material goods than ever, whilst arguably being more dissatisfied and insecure than ever.

There is a 'civil war' going on inside many people – an inner stress of tearing contradictions and despair. This is fuelled by living with great fears, inflated expectations and an erosion of self-belief, self-respect and meaning. This overspills and plays itself out in conflicts across this planet, from the crime in your neighbourhood to the shifting front lines in warzones, many of which go unreported.

Such a speed of life, speed of talk, speed of consumption. A human race racing fast, frustrated, angry, overheated. Does this actual 'human warming' not relate directly to global warming? So when we witness the melting of the polar ice caps for example, aren't we actually witnessing the unfolding and direct consequence of the change of climate that has gone on in us, because of the lifestyles we now demand that are costing the earth?

These are first principle starting places to understand the origination of some of the key aspects of climate change, and where remedy could naturally arise. Yet trying to be remedial as a first principle can only yield limited, short-term results at best. The answers never lie where the problem is.

A more telling definition of the word *climate* is *prevailing attitudes and standards.* These are currently in great flux, along with the prevailing priorities and views about how people can do more than just survive into the future.

So as a first principle, climate change is happening inside the human now, inside humanity, inside you.

In which the tiny universe of our own teeming internal energies is directly related to evolutionary changes and tides incoming and outgoing within the infinitely bigger climate of the universe we live in.

Creating your own climate

Each person creates their own climate with its own atmosphere. Perhaps you feel that yours is a 'high pressure' climate, or that you are too often 'under the weather'? Or maybe yours is a still, warm, gentle climate inside, or sometimes too cold, bleak and overcast?

Your internal speeds and balances in the way you live your life can create a welcoming or hostile climate. So the hasty, frenetic and anxious way in which we can find ourselves experiencing life today can trigger driving pressures that make for a stormy and turbulent inner climate.

This in turn can create a widespread background atmosphere of frustration, hostility and even aggression in oneself, one's home, one's workplace. And when such a hostile climate interacts adversely with others, it can only spread and strengthen, harming people and the environments in which they live.

Within this, the more powered situations arise where groups of people end up generating a climate of fear, panic or revenge or, on the other hand, peace and hope. All these climates originate from and directly affect human behaviour and what we do to each other and to the planet.

Humans decide. But if people decide about the value of the planet without considering the value of human life, or vice versa, then we could end up losing both; in which the value begins with a mindful appreciation of and respect for all life – universal, human, planetary, fauna, flora – and how each part is totally interconnected and inter-supportive, by design.

A change of mindset

If we speak of natural climate change in the human we are speaking principally about a development and change of mindset; a change in the way we think, contemplate, feel and perceive. And this is one of the pivotal points...

... because a hundred new laws to tackle planetary climate change can be brought in, but without a change of mindset, there won't be enough will or value in people to see it through. New laws can be laid down ordering people to recycle more as an example, but when people feel the precious value for life and for the planet that is their home, they'll be more likely to do it of their own volition – and much more.

Is this naïve or too simplistic? Well the evidence is that almost all the complicated and sophisticated solutions have not significantly changed either the human climate or the planet's, because they rarely question deeply enough or go to the core and causes.

So there is an urgent need to facilitate a change of mindset to one that could understand and deal with climate change for real; a change of mindset that moves closer to the natural purpose of human life and the purpose of the planet, by appreciating, studying and respecting the true natures of both.

A clear, constructive mindset can evolve, by reasoning, logic and agreement, to take some positions as to what we as a global tribe of human beings are aiming to be and do with our lives; within which to clearly decide whether we want to do this at the expense of humanity and the planet, or not.

The case for urgent action where the planet's climate is concerned has been self-evident for far too long, but only through understanding and affecting change in the climate of our own lives as people can lasting change take place.

Change and alteration

One of the major blind spots in all this is to do with the amount of time and effort invested in 'alteration' thinking and actions which rarely effect real change. Alteration thinking leads to policies such as reductions in carbon emissions or increases in renewable energy by conveniently small percentages or amounts that will be politically expedient; whilst the truth about the magnitude of actual change needed is often hidden or avoided.

A small percentage increase in recycling or fuel-efficient cars or solar panels and suchlike, may be a good *alteration* in behaviour, but not nearly enough to effect the actual *change* needed nationally or globally.

In many cases even wide-ranging alterations can make matters worse, by creating the illusion that they are ultimate solutions. For example, some renewable energy or pollution-reducing technologies – such as wind farms or catalytic converters – in their material construction, maintenance and output, can generate more carbon or pollutants than they reduce.

There is a massive difference between alteration and change.

In the context of the planet's climate, many well-intended efforts were made as an example to collectively reduce greenhouse gas emissions by 2012, by 5.2% from 1990 levels (Kyoto Protocol, 1997). Limited progress was made, whilst many expert climatologists had already assessed the need for far greater reductions.

In this example, a 5.2% reduction is hopeful alteration thinking, not real change. It means a more comfortable and palatable transition that may slightly improve things in the short-term, but will be irrelevant for future generations who will face an increasingly barren planet.

The Kyoto Protocol was the first major international step towards seriously tackling planetary climate change. Yet a few years later, some scientific projections were already showing that the Kyoto agreements would probably only reduce rising temperatures by around 0.07C by 2050.

This is not a change but an almost irrelevant alteration when you consider that current research shows that temperatures could rise alarmingly from pre-industrial levels by up to 2C or beyond by the year 2050 (IPCC reports and others). The impact of a rise of just a few degrees on this planet could be, without exaggeration, catastrophic; we could be struggling to even survive on a seriously overheated planet, one third of which could become desert.

Agreeing a major reduction in greenhouse gas emissions now would challenge the overall climate of comfortability, fear and casual expert views, as well as the way we think about ourselves as human beings and this living planet that graces and feeds us. It would challenge our sense of humanity and perhaps reflect back to us how far we have drifted from a purposeful and connected way of life.

It may be easy to roll out the thinking, reasoning or policies of alteration from the comfort of an executive chair, not worrying about whether you will have enough food or drink today.

But if you live in a poor region seriously affected by climate change, with parched land and a polluted water source, not knowing if you will survive another week, then truth, reality and change become your only hope.

Deforestation, as another example, is already directly harming millions of people across this planet, as well as her flora and fauna life, every day. We are literally starving future generations of the oxygen to even breathe, let alone leaving them enough food to eat.

The criticality of this however originates as a first principle from the way we have chosen to lead our lives today; it originates from the way we think about ourselves, each other, and our greater purpose in being alive on this earth; it originates from the way we may tend to undervalue ourselves, other people and this planet, or inflate our self-importance.

This ultimately pollutes the human race every day more powerfully than the build-up of greenhouse gases, which in reality are an end result of such a way of going on. Yet is this a perception that drives and focuses our minds and actions?

Change and sustainable development

Change is entirely natural in human development. We are designed for it, in the same way that seasons and stars are designed to change over time.

But the climate that people are brought up in, in industrialised nations, is mostly a toxic one of exaggerated and unrealistic wants, demands and expectations that places the human at the selfish centre. This cultures a polluted climate of imbalance or dis-ease and inhibits change, whilst depleting the planet's ability to heal herself.

And it turns some of the promise about sustainable development into sophisticated alteration talk. What can truly be sustainable when at the current rate of population growth and industrialisation we've already rendered the planet unable to sustain us now, let alone for future generations?

Do we, as a global population sharing one planet, honestly believe that the current levels of extraction of oil, carbon, gas, metals and minerals are in any way sustainable, or that the frightening levels of famine and disease and warfare are tolerable?

What kind of a world are we planning to leave behind us when we depart this once unviolated planet? Can we rediscover our humanity?

Sustainable development applied to the planet could be remedial, but only if partnered with the development of a personal and collective will to engage in the education and changes needed in human behaviour – otherwise sustainable development becomes the accepted panacea and builds in its wake a convincing, alluring rhetoric to pat all our problems into a neat unsustainable shape.

Sustainable development as a first principle is about a human's ability to nurture a humanity, a consciousness and a conscience that frees us to be true, compassionate, mindful human beings.

For technology will indeed save us, but only the human technology inside us, of questioning, listening, understanding and facilitating real change for good. This is a lifetime's ongoing personal development, and yields the kind of sustainability that will make all the difference, now and for the future.

Climate changes in humans

So are we content to just travel hopefully through this current deadly climate of human dissemblance, holding tightly to the fantasy that we will eventually cure all disease with drugs, conquer terror with terror and solve pollution with technology, thus allowing us to consume the last remnants of the planet's bounty? It truly is insane.

Many of the clever short-term answers are making a big noise of confusion to obfuscate the real questions to be explored – such as *what does it really mean to be and act human in these times?* Can such questions be heard and pursued? Or will the quick-fix solutions drown out the instincts and cries of the children of a new generation who may never understand how we were able to cause so much wholesale destruction of creational, human and planetary life in less than two centuries?

What if we were to look through young people's eyes and see the reality of the world that they have entered?

Are we surprised for example at the levels of drug abuse that begin at such young ages? Is it that some would rather become numbed to the profound sense of hopelessness and fear they instinctively feel, than face a future they can no longer believe in... with few leaders they can look up to and trust?

If we are serious about this, and many people clearly are, we will need to discover, understand and encourage the climate change that needs to go on inside each of us. Otherwise all efforts may simply prove ineffective, and the beauty and purpose of being human, and the wonder and grace of this extraordinary planet, may truly be lost forever. The stakes couldn't be higher.

We are spiritual beings first

Perhaps in our calmer, more reflective moments, we truly feel that we are spiritual beings as a first principle, seeking human experience as a second principle.

Yet sometimes our human experiences in life, which may harden us or make us bitter and unforgiving, can suppress our spiritual nature. Holding the essence of bitterness or resentment is toxic and harmful to our own systems as well as to other people; it carries our history forward as a burden rather than a learning. Experience, good or bad, is an education from which to draw wisdom, to improve. Experience is what we have to liberate and enhance our essentially spiritual nature.

And it is from this benign nature that we treat ourselves and each other as humans with being and create a fine, changing, progressive climate of humanity between us.

This is realistic, not idealistic. For fighting against our inherent nature only weakens and harms us, whilst embracing it sets free the longing we each have to be ourselves. From this climate, everything is possible.

Which natural human qualities would you
consciously want to grow in yourself?
How would you begin this?

*

What would you want to improve in the climate
of your behaviour towards yourself and others?

*

How can the sacredness and joy of life be
rediscovered?

Liberating perceptions

The search for truth

May we see the great lightworks
of those rhythms and reasons
for why life is given
as they spark the quiet skies
of our deepest reflections

We, a tiny universe
of moons, planets, stars
evolving and revolving
around the motives
that guide our choices

Our freedom is a true response away
or an addictive illusion
if self-gain grips the life from us

Truth only hurts us
when we are far from it
or we run from its message
which is to search for it
never deny it
love it
realise it
live by it

A gateway of contemplation

If we knew the truth about the kind of person we have become, would we accept it, want to discover more, search out what to let go of?

If we knew the truth about why other people are the way they are, would we change our attitude towards them?

If we knew the truth about the planet's feelings, would we change the way we feel about her and treat her?

A gateway of reasoning

What is truth?

In the recent past, experts told us that commercial fertilisers, insecticides and herbicides were crucial for farming, but today we are counting the cost. In the past manufacturers promoted asbestos as a safe and effective building material, but today it is banned almost everywhere. One year a particular diet is hailed as essential, the next year it is discredited.

Somehow we can get addicted to the apparent truth of the moment, a temporary truth perhaps. Sometimes these may reveal a permanent or deeper truth, but more often they turn out to be hollow. A new concept, fad or hyped-up medical discovery can prematurely be seized upon as an ultimate truth, often before anything has clearly been proven.

So if we are to begin the search for truth, we need to find and discern permanent, lasting truths which are not subject to passing trends, static opinions, theories or people's likes and dislikes. For the truth endures and has no bias.

For example, it is a truth that we become what we think about. Our thoughts govern and shape our behaviour and character over the course of our entire lives – they are what we become.

It is a truth that the more dogmatic our reasoning becomes, the less open we are to what we don't know. It is a truth that each sequence of a person's actions in life carries a consequence. It is a truth that we cannot and are not meant to know everything about the universe we are in or all of what awaits beyond our earthly lives.

Life pulsates with truths, in the causes and reasons for what exists. The truth of a rose lies in what caused such beauty and for what purpose – in what a rose is unto itself, irrespective of what we may personally feel or think about it. To us a rose may be a beautiful addition to a garden, but for the planet, the air and the bees it is much more.

By the same principle, the truth of a person lies in what caused their existence, and in the motives and ways of their response to the life they've been granted.

We connect to truth when we realise it. Then it may grow inside us, and we become nourished by the fruit of its wisdoms. These guide us to stay aligned and true to what we are and to what gives us life. For we are designed to seek and know truth, and to let its pathways lead us away from self-deception and into the light of what works.

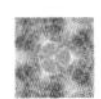

So truth itself is a guide to live by, in the same way that an accurate map provides true guidance. Yet we often need to travel through many fields in life to discover a truth. And that whole journey of seeking, discovering and realising truth can take time. Think how long it can take to discover a truth about one's deeper nature, longings, blind spots or natural inclinations. And think how long it takes to discover a truth about what caused the Ancient Egyptians, Babylonians or Sumerians to think and live the way they did.

To seek truth is to venture beyond the end results or facts about anything, whilst perceiving where they lead. *For facts are end results, whereas truth lies in the causes.*

Whatever a person speaks out loud for example, is an end result, a fact in itself, you hear their words; but the truth lies in their reasons for saying them.

Similarly, it is a fact that we will die one day, but there are truths about the meaning of death and the art of dying that require more than factual analysis.

There are many facts about the Ancient Egyptian sarcophagi and tombs that we know of, but what of the truth that lies in their reasons for creating such elaborate and ornate funerary artefacts and ceremonies in the first place? What truths were held in their minds that caused them to build the pyramids?

The search for truth is a journey to the core, the cause. It's getting under the skin or surface display or outward appearances of something. And it takes us into the natural laws and patterns everywhere in nature; into the deeper truths behind the fact of dawn and dusk, the mystery of birth, the meaning of sleep and dreams, the healing forces in colour, the power of posture and movement...

For all these originate in a supreme truth, that *everything begins in the unseen worlds* of causes, reasons and purposes. These are the realms of intelligence and power governed by natural laws. These are true, for nature doesn't lie or try to be something it is not.

In which the high causes and indelible pathways of life that nature expresses, give permission and possibility. They are faithful, constant, liberating and true to their purpose. And when we travel them, we come to know more of the truth of our own existence.

In touching such universal truth we touch some of the mystery that lies behind the enduring love that grows in and between people, or the devotion of parents for their children, or the inherent seeking to understand why things are the way they are.

Universal truth reveals that life is an infinite, interconnected, sensitive weave of permissions and possibilities, in which each human can play a conscious part in its upholdance and refinement.

For the truth is that we are instruments of evolution. Thus our birthright is as brothers and sisters to a high cause.

Each human life is born from the same blueprint and yet is completely unique and original at the same time. When we celebrate these differences we discover another truth: that no one person can hold the whole truth, and that we are all part of humanity, designed to work together as world citizens, when we strive for something greater than personal or national ambition.

In which the allowance and tolerance and encouragement of different expressions comes with living the truth that humans are diverse, evolving, trial and error, unfinished.

Then we can seize a further truth: that we are designed to make mistakes, to learn, grow and refine by them, and to not condemn ourselves or others for making them.

A gateway into the unknown

When we look at our face in the mirror, or at the faces of other people, do we see a truth or only the outside mask of a truth? Are we only seeing what we want to see as we look through our eyes – clouded as they may sometimes become by familiarity and habit?

Each line, tension and expression on our faces is an exact portrait of a truth. The truth of what we care for, are working for, long for, fear for and worry about. This leaves its trace upon our faces and the lines on the palms of our hands, as an engraving of what we have made of our lives.

Yet beyond this is a deeper truth. A truth that speaks in feelings and motives. This is the truth of what and who we really are, beyond the body, and what we draw our motivation from.

And the truth is, so much of what we are is completely unseen and unknown. We may hardly know our own instinct, our own spirit, our own mentality – all these majestic, powerful faculties are scarcely explored.

We are indeed mostly unknown to ourselves, like the universe we live in, like the universe of our mind, like the universe of our own DNA.

In which the question arises: does a person want to explore the real unknown of life, including their own life, or remain fixed in the confines of what one already knows and can do?

And what happens if a searching question such as this challenges our comfortability? Will you or I be pleased that a question can take us into the unknown and melt our self-limitations?

Many people say that they really want the truth, but what if that means years of working towards greater openness, self-discipline, perception, developing new skills and constantly renewing one's positions and views each day... would we still want the truth? And what if the search for truth also means discovering more of our impediments – like being apathetic, manipulative, quick-tempered or a mental bully – would we still be keen and willing to discover more?

We each have a self-view. But what if this view is too narrow, out of date, too superior, too inferior? Would we still want to know the truth about it? And if we did, how could we find a better way to proceed, without blaming anyone, including ourselves?

Those who seek truth seem to actively find themselves asking fundamental questions all the time, such as: *what is causing this? what does that mean? why did I do that? what did I learn from that? how could I improve?*

The fact is that we are as we are today, but the truth is in why that is the case.

If truth lies in what causes all life, including human life, then the seeking, loving and living of truth draws us closer to our very own nature.

What does being true to yourself mean?

We lack nothing
but the perception to see what we're made for

The need to perceive

How is it that with so much more information available today, there seems to be less understanding? How is it that the more we use computers and phones, the less we seem to truly communicate at depth? How is it that we know about the daily tragedies in this world in much more detail now, but are less effective at being curative?

Is it that we fail to perceive the reality of our situation as it actually is?

The apprehension and noise of these times cloud our perception. They distance us from ourselves, causing us to sometimes be beside ourselves, as opposed to being inside ourselves; or driving us out of our mind as opposed to being in our right mind. The natural harmonies and fine balances that can settle at the centre of a human can so easily become stressed and fragmented, putting us on edge and out of sorts.

Perception is the core, innate capability and process that builds the presence of mind to see the truth of what's really going on. It brings us back to our senses.

We may hear many words, but we only perceive what they mean when we're truly listening. And our thoughts only reveal their meanings when we perceive what causes us to think the way we do.

Perception is a worked-for freedom, to understand at depth. It is a freedom born of the kind of self-development pursuit that enables a person to find a route from arrogance to humility, from self-centred to self-discovering, from the predictable to the unknown, from everything being personal to everything being possible.

Examples abound...

A surprise discovery reported in recent times was that complex cranial surgery was carried out in Europe around a thousand years ago, and much earlier in the Middle East – obviously without modern anaesthetics. In fact many ancient civilisations and tribes had advanced healing procedures and practices, working with energies, colour, sound and herbs as well as surgery.

Such practices came from perceptions about illness that focused on the causes of internal imbalance, dis-ease and disharmony rather than merely outward symptoms. In this respect healing would have been perceived more in terms of maintaining a constant inner harmony as a first principle.

As another example, the Chart of Dendera from the Temple of Hathor in Egypt presents a map of the heavens as they were perceived over two thousand years ago, giving great scientific and intuitive insights into the zodiac constellations. Those same starfields may be more visible today through the telescopic lens, but the perceptive mind offers deeper magnification and meaning, as can be perceived on the Chart.

There are many other examples of advanced accomplishments in the human story over the millenia: the creating of languages, fine arts, the building of pyramids and the achieving of a high and connected development that understood so much about life and the afterlife. All this through awakened minds, rather than a purely mechanised analysis.

Meanwhile, science has opened up a lot more knowledge today in so many areas. Yet even with masses of data and new discoveries, perception still eludes us in so much. Earthquakes for example have been researched extensively, yet we still seem unable to perceive why animals often detect, 'dance' and react to earthquakes some time before experts read the signs on their instruments.

Is it because animals are connected to this planet more directly than most humans today, and can therefore naturally sense radical disturbances in frequencies, sonic waves and rhythms in the planet – to which they instinctively react and flee?

Such connections can happen powerfully between people who may be at distance, but are 'on the same frequency' or wavelength and can thus tune into events sometimes many thousands of miles away. This is also how intelligence transfers and networks, and how original discoveries are made simultaneously in completely different parts of the world.

This natural telepathy is true 'wireless' communication, and is possible when our minds develop a finer ability to transmit, receive and perceive. This happens with many people every day in small ways, for example when you think of someone or picture them in your mind, and then seconds later they make contact. A firm belief in our innate ability to do this is an important start. From then on it's exercising that ability in multiple ways, without censoring oneself.

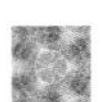

We live in times when almost all the foundations of current human organisation are breaking down: systems of government, leadership, society, finance, industry and economics are mostly in dangerous freefall. And through this, the planet's sensitive ecosystems, ozone layer and natural self-healing mechanisms are also breaking down. But do we perceive why?

Have we become enslaved to a daily pattern of action and reaction wherein we rarely find the right speed, space, stillness and openness to truly perceive what is happening and how best to respond? Think of the emotive reactions to tragedies in the world, which often fuel more of the same – such as when a reaction of revenge and attack quickly overrides humanity and understanding.

Decisions made with the easement and calm of forbearance, forgiveness and foresight however allow much greater perception, and produce a more exact response to a need.

Maybe it is very naïve to hope that anything can fundamentally change the collision course that the human race has been taking, especially since the explosion of industrialisation and what fuels it.

Perception can.

It begins when there is the awareness to be able to pause, to reflect, to admit that perhaps we don't know, and to slow down enough for understanding to find a fertile place within us.

This can begin in small moments, when we are able to deliberately stop ourselves at the right time, during the busyness of too much to do, to remind ourselves of what is most important in life, and why.

At such times it becomes possible to elevate in oneself from thinking and reacting, to perceiving and responding. From this elevation, it is natural to perceive both the sequences and the consequences of our thoughts and actions, enabling us to be more mindful from a long view and thus more effective in the short-term demands of living day to day.

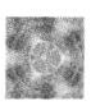

Perception is the art of original thought, uncommon sense, listening to the instinct, being more fluent in the obvious and the esoteric... and open to serendipity. It is in having the awe and appreciation for life as one's motivating discovery urge. It is an innate need we feel as human beings to know and respond better.

Perception is not born of the insecurity of adopting an intellectual persona or style to be applauded by others. It is born of a deep-set longing in each person to be a little more wise, caring and responsive today than we were yesterday.

To perceive is to transcend the fear of the unknown and embrace it, especially in oneself. It is to rediscover that the purpose of the human mind is to enquire, to wonder, to create, not to destroy or stagnate.

And many minds joining together on such a journey can become a true catalyst for change.

This is not a fanciful ideal. It is a natural journey called life, and we have all been gifted one to live.

All the signs of these times we are living through – in both the extreme degeneration and the nascent regeneration – are crying out for us to perceive our situation as human beings more deeply, so that we can respond responsibly rather than react ineffectually.

We may perceive what it takes to survive in the twenty-first century, but do we perceive that this life, here inside us, is the start of a universal life?

What does it mean to be human?

Reincarnation, bling and your greatest mistake ever?

In the great striving towards the freedom and fulfilment that life holds in store, it becomes important to try to understand how apparently unconnected things may be significantly linked together. And the three subjects in the title tell a significant story, when you make the links.

So this essay attempts to open up and describe each of the three subjects as three interlocking key pieces from the bigger puzzle of life. When you see each piece described and coloured in, the links will hopefully become clearer as they join up to reveal a picture perhaps not seen before.

But firstly, why would it be important to see the connections between such seemingly disparate subjects?

Making connections

There is such a magnificent multiplicity in life, and part of the art of engaging full-heartedly with it is to perceive both the obvious and subtle threads of inter-connection that exist.

There are always different levels of understanding about most issues, as life moves on all the time inside an infinite universe with its own levels: galaxies, constellations, suns, planets and moons all exist at different levels in the cosmos, with ranges of energy and matter, high and low, fine and coarse. Whether we can perceive it or not, all things occupy their own space, domain and function, whether they add to or subtract from the great purpose of it all.

It is 'as above so below' and also as above *with* below. For we can watch the skies and admire their beauty, whilst at the same time carrying that same beauty inside us in the atoms, chemicals, gases, light, space and intelligence that we are each made up of. In this sense we breathe in the sky each moment, we live in it, in direct connection.

The infinite traces of life connect up in ways seen and unseen, from the heavens to a virus. All living things have a sequence and a consequence, a cause and effect, a rhyme and a reason.

A heated confontation at work and the killing of a soldier in war may seem disconnected, but they both involve aggression of some kind. Even a single act of violence, on one street in one town in one moment, can trigger years of war right across the planet, as was witnessed in the outbreak of World War I.

As another example, in the West, black is traditionally worn in funeral ceremonies, whereas in China different family members wear different colours. These are two contrasting expressions, but they are both linked by the shared values of respecting and recognising the death of a person.

Working out and understanding such weaves of connections is a powerful personal development exercise. It nurtures the ability to discover greater context and signficances, and therefore greater appreciation, understanding and response.

It is a detection of what fits with what and why, and how to then best proceed.

You may live on a remote dirt track in the middle of nowhere, but when you see how that track may connect with the whole nation's network of routes you are then free and able to travel far and wide.

In making connections, we can also enjoy discovering how the sublime lives a breath away from the ridiculous, and the ordinary a hair's breadth away from the extraordinary.

And so whilst this is a serious journey to make, it will not proceed very far if one takes oneself too seriously. Taking yourself a bit more lightly enables you to travel more lightly... less weighed down by the excess psychological baggage of your past or current troubles.

Hence this delicious, eclectic medley of the three seemingly unconnected subjects in the title, which blend the higher, middle and lower aspects of living – all three levels being crucial in the holistic development of all that one is.

For our individual lives are created to connect with all elevations of life. And the quality and level of our journey through life on earth and beyond depends on what we are open to, what we love, what we strive for and ultimately, what we become connected to. We are truly never alone.

So here are those three pieces of the puzzle which fit together, when you stand back and look at the bigger picture that they make...

Reincarnation

A first level understanding of reincarnation has a simple logic. Everything in nature works in cycles: birth, growth and death and birth, growth and death... animals, trees, flowers, insects and even clouds and seasons. Life is a constant coming and going.

Human beings at one level are just the same – child, adult, parent and child, adult, parent and on and on through the generations. And if you look at the individual, this also works by the same principle. The energy that a person constitutes can change its state and function, like ice turns to water to steam and back again.

As a physical body, humans are made up of bones, nerves, blood, organs, etc. But as a whole person, we are a bundle of teeming energies in terms of our thoughts, memories, knowledge and many qualities and senses. Through sleep and food these are replenished each day. To do what? Well, to invest in some way in being and doing what we believe is best in life to do.

And what happens to that energy at the end of our planetary lives? In very basic terms, it depends on what potency and quality of energy, force and power our lives have become, through the sum of our endeavours.

In which there are three levels to look at, impersonally, without morality or the judgement of right or wrong:

Expendable – energy that is low power, cannot hold, is dispersed

Maintenance – energy that is nutritious, good quality, sustaining, useful

Custodial – high, fine, lasting, generative, creative energy, force, power

Expendable energy eventually disperses. It's not destroyed, but cannot hold together as a continuing usefulness or function or capability. It's a bit like blowing up a balloon which then bursts – the air or energy just disperses back into circulation and the balloon has no further function.

Maintenance energy has a possibility of continuance, because it has vitamins and strengths and a common purpose that binds it and lasts. This is where the development of human qualities and values becomes so important in a lifetime, because these are powerful forces that have a radiant energy life that can endure. This is sometimes recognised as being a fine charisma and aura around a person, which makes others feel well and at ease with them.

This level may yield a reincarnation possibility, where the energy store that a human has become at the end of their life carries a power, strength and quality – and thus a further usefulness – that may allow the possibility of continuance after death, in a new body or form on earth.

It's in the word:

Re – in repetition
incarnat(e) – in the flesh, given a bodily form
ion – with charge, power, force

So reincarnation lies within the range of human possibility but is not automatic. And the prospect of reincarnation, as attested to by many religious and spiritual teachings over thousands of years, is the prospect of further life on earth.

Custodial is a level beyond reincarnation, when a human connects with and is custodial to influences and powers above the planetary domain. This is based on the endeavour and perception of oneself as a spiritual being, whose home and purpose and function lie within the higher creational and universal domains. This can become a quest or movement towards or beyond immortality, and into the multiple realms of service, being, and co-creating.

The principle at play is that if what we become consciously connected to during our lifetime lasts timelessly beyond the physical, then so may we.

Is this not a matter of how well or how finely we educate, assemble, align and 'dress' ourselves, from the inside out?

Bling

One simple, wise philosophy in life is, 'don't become too attached to things – you can't take them with you when you die!' This doesn't only apply to cars, houses and your favourite item of jewellery, but also to your opinions, judgements, habits, identity and your own body. These are all left behind.

Therefore, are our latest disappointments and resentments, or gadgets and self-adornments, the kind of things we want to become too attached to?

If a person knows that they are not their body but their consciousness – in the sum of their knowings, feelings, principles and all that these are connected to – then this may cause them to think twice before becoming too possessive about 'my this or that'. If you are not your body, you are certainly not your car, home or jewellery – even if you love them!

However, if you see your whole character to be defined by such possessions, then they become what you are, your identity. In this respect, what you possess possesses you.

If you own a cherished item and it gets stolen, will you be able to get over it? If yes, then your life is perhaps not governed by the ownership of objects, and consequently, you may enjoy material things a lot more – you may own things, but they don't own you. It's a bit like earning heaps of money so as to live in a huge house with bars on all the windows – do you own it or does it own you?

The joy, laughter, sadness and tears, along with the physicality of life, are all important, but are ultimately expendable... leaving the question as to whether we want the whole of our lives attached to that which does not last?

It's a matter of *what you want*, which is the most fundamental starter question of all in personal development. What do you want?

Our behaviour in today's world is often conditioned from an early age with imbalances in our driving desires, aspirations and demands. This may instil an ongoing insecurity that is only temporarily relieved by attachments to things that seem to make us feel secure and worth something.

The wearing of designer labels proudly displayed in symbols and patterns on clothing can make us identify with the charisma or class of that brand and label, and leave us feeling good, superior, whatever.

Fine, but if we value ourselves according to these labels, or even labels such as good-looking, good competitor, clever, then…

In a similar way, the big emotional highs of 'I absolutely love my new computer/ outfit/watch' may leave us somewhat depleted when it comes to truly feeling love for oneself or another human being.

If we could afford it, we might dress ourselves head to toe in the 'bling' of fine gold, silver and shiny diamonds on the outside, but can we match such beauty with a fineness of character and illumination on the inside?

Your greatest mistake ever?

How do you see yourself, and how do you respond when people ask, 'So what do you do?'

Well, if you ask a range of people they may say, 'I'm a secretary. I'm a famous musician. I'm a primary school teacher. I'm a student aspiring to be a journalist. I'm a father. I'm a consultant surgeon. I'm an atheist, I'm an environmental activist.' And of course, all this may be true.

But is that it? To what extent do we just become a 'one thing', characterised by one word, one job description, one station in life or one set of opinions to the left, right or centre? Is our self-view the limiting one we may have ended up with, or the unbounded and diverse one that is possible?

Everything about the way a human being is designed speaks of multiplicity: many different senses, many faculties, many internal systems, many potentialities, natures, inclinations, arts…

Often from a young age, however, we are already being coaxed down a one-way street: 'I think he'll be a great business leader when he's older' or 'She'll be an amazing mother when she grows up' and such like. All this plays on our minds, together with the fact that as we live in a world ruled by economic criteria, we are pressured into making up our minds and matching these criteria sooner rather than later. This sometimes prematurely restricts the fuller potential of the whole person.

Even if we *could* all fulfil our career dreams, these days, according to recent studies, this would mean that we'd have millions more TV presenters, pop stars and celebrities, with fewer people left to bake the bread, teach the young, or offer new principled leadership into the world.

The truth is that each person is multiple, in the potential range of their skills, passions, inclinations and aptitudes. When compelled to identify ourselves with just one or only a few categories, such as 'I am a good builder/thinker/swimmer' or whatever, this can hugely shrink down our self-view and potential. And this can eventually become oppressive upon our natural spirit and human instinct to be open to the greater multiplicity of life, which inspires us to explore, to discover, to search into the unknown, to see what else we can become.

Life today may be about becoming a specialist or a recognised 'something' in society, to gain status and wealth, but this pressure and/or enjoyment does not need to leave us inside the box of only being that something and nothing else. It is a choice, but who or what is making that choice?

For who knows... a person may not allow themselves to think so, but given the chance they may discover that they are more outgoing, gregarious, reflective, bold, sensitive, compassionate, witty or artistic than they ever gave credence to.

And from this, they may also realise an unknown desire to try calligraphy, juggling, geology, playing a musical instrument, country walks, meditation, astrology – a million things they may never have tried or even imagined trying. The range is unlimited, the choices inspiring, the feelings extraordinary.

So limiting yourself, your options and your freedom of expression, by trying to handle the whole of life through one fixed self-view, could become your greatest mistake ever.

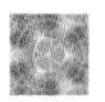

In now looking at the bigger picture that these three pieces of the puzzle make together, there is this reflection to offer:

The wearing of jewellery, medals, titles, identities and status may feel good, bring good and make a great impression on others.

In the end though, the constant, genuine, bright efforts we make each day, and the fine qualities we nourish, become the real, precious gems that crystallise within us. They show our many true colours, and reflect the light of human kindness. They are the treasures that last forever in our quest to be useful.

Do you see yourself to be a human
seeking spiritual experience?

*

Or do you see yourself to be a spiritual being
seeking human experience?

*

What difference does it make?

Hardness can support softness,
in new growth

It can be a significant and fascinating process, to explore a certain word that is commonly understood in one way, perhaps pejoratively, to then discover that the truth is very different. Like the meaning of the word...

Hard

The essence of hardness in a person carries a natural quality of being firm, resolved and steadfast in what they stand for. This nature of hard provides strength, integrity and safety – it fuels the will to go on. This involves standing on principle, yet being open to change. It involves being honourable, yet willing to review what that means.

In essence, being hard is about being reliable, true and constant, with great endurance and mental toughness – hard like in hardy. Dictionaries feature this essential nature of hard when they define it as: *very potent*, *powerful*, or *solid*, *firm*, *intense*, or (of information) *reliable*, *especially based on something true or substantiated.*

This is not the harmful, abrasive, aggressive hard, as is often characterised in human behaviour, but a natural hard, in some ways similar to the character of hardwood trees.

Hardwood is obviously more durable than softwood in building, because of its inherent nature and function. You would use hardwood for the front door and softwood for the shelving indoors – each simply fulfils a different need.

In this natural sense, hard is not in opposition to soft, for they are often integral to the same process. You may be soft in your listening and openness to the needs of a friend in trouble, and hard in the sense of reliable, firm and upfront in your response. This kind of hard has the character of steady, resolved, responsible, and works together with the being soft, which is acquiescing to whatever is called for. Neither state is good nor bad, it's simply finding what approach works with what situation... usually a blend of both.

The natural development of hardness in the course of living can be characterised by the phrase, 'the mettle has been tested'; in which your experience in life can harden you in a natural way, to be strong and resilient enough to follow a principled and self-determined way of life.

This is analogous to metal being heated to the right temperature, softened, moulded into the tool or shape required and then left to cool, set and harden to retain that shape and character. It's now ready for use. And it's the same with growing tomatoes from seed: as the shoots become strong enough in the greenhouse, they can gradually, day by day, be introduced to the climate outside – a process which is actually called 'hardening off'.

In how we think about all this, we can say to ourselves for example, 'It's hard to train to become a doctor.' But it depends on what is meant. Training to become a doctor simply takes what it takes – many years of dedication and intensive study and practice. It may be hard in the sense of challenging, but that nature of hardness is simply part of the journey for someone who wants to undertake it. That training can be soft at the same time, in the sense that practising medicine can provide much well-being for others and much satisfaction for the practitioner.

So natural hardness is not inflexible or brittle. It is simply the firmed-up state that it needs to be. Even our own bones are not *hard* hard, but *porous* hard, with an inner softness in bone marrow and an ability to regenerate.

And if the 'h' in hard stands for 'human', and the suffix '-ard' stands for 'ardent' and 'ardour', then this reveals another aspect of the natural character of hardness in a human.

All this is very different from the misplaced hardness in human behaviour which is reactive, aggressive, bullying, forcing a point of view or oneself upon another, being dogmatic or unforgiving.

This kind of hard has no place in a naturally developing life, and no place between people. Being hard, in this type of human behaviour, is being out of character, oppressive, critical and harmful upon oneself and others. It is against the grain of the natural growth patterns in human behaviour which are inherently soft, flexible, malleable and open to change.

So much about our modern world has an unnatural hardness, in sharp edges, tight corners and boxed rooms; in noise, anger and rage; in the overuse of concrete and harsh neon lights; in grating sounds, clashing colours, punishing words.

This is antithetical to life and to the very essence of the human design, and is to do with what is hard that should not be hard.

We may say to another person, 'I think you are being too hard on yourself', but at the same time, do we stop to consider if we are being too hard a taskmaster upon ourselves, or others who don't match up to our expectations?

One of the finest personal developments that one can make in this whole arena is the nurturing of restraint: the quality and ability to hold oneself back at the point as needed, and to stay calm, reasoned and not overreactive when something really challenging flares up.

This is a hard discipline to acquire, because we all get influenced and conditioned to react without restraint or reason in difficult situations.

Restraint works like the anti-lock braking system (ABS) in modern cars, which prevents wheels from locking and thus cars from skidding out of control. When you have to brake suddenly in an emergency, these brakes, instead of gripping with immediate effect, actually cushion the braking through a rapidly opening, closing and pulsing valve and pump system. A typical ABS, when activated, can apply and release braking pressure at up to twenty times a second. Now that's well and truly practising restraint!

This is a very telling example, because when we slam the brakes on too hard in our lives, or indeed accelerate too fast for the journey we are on, we need the cool head of restraint in order to keep our aims and priorities on track. It's a fine balance to find in the blend of soft and hard.

Another example in life is to do with making agreements. Often people try to be too decisive too quickly about important agreements to be made – effectively, too hard-and-fast, or too hardheaded.

It's true that an important agreement may be reached promptly and with ease, but usually there are many considerations to reflect upon in respect of the aims, terms, consequences, details and timings. Then there are all the considerations regarding the upholding and keeping of that agreement.

So a natural way to proceed might be:

- To approach an important agreement softly, in a timely manner, with care and conscientiousness; let the agreement form up with patience and with space for the needs and wants of all parties.

- Then, when the agreement is made, it needs to be upheld honourably and with commitment – a hardness that also contains an openness to review as needs change.

In making agreements, it's important to watch for the balances, timings and flexibilities needed in the movement from soft to hard, and perhaps to soft again, because circumstances change, even when the intention does not.

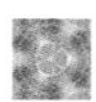

Life is hard and life is soft – there's no point in getting personal about it. For each state can become the necessary resistance or easement in developing the best ways to find what you're looking for.

In what ways are you hard?

*

Which ways work, and which harm?

Let the needs lead

One of the most crucial needs in this twenty-first century is a new kind of leadership. A leadership that lets the needs of any situation lead – always with a long view of the generations to come. A leadership which understands that leading by living example is the essence of all natural leadership.

The leadership spoken of here is not about people having authority over others. It is speaking firstly of self-leadership, in changing times, where a person becomes willing, prepared and happy enough to take both the responsibility and the consequence of how they choose to lead their lives.

Self-leadership is a development journey towards knowing yourself. It is becoming conscious of and refining the talent and attitude and substance of what one has to contribute as a life. It is also an ongoing awareness of one's weaknesses and blind spots.

The first signs of such development are to be found more in a person knowing what they *won't* do, having worked on what their principles and standards are. This builds trust and self-respect. From this platform comes the development of an ongoing openness, equanimity and willingness to listen and learn, whilst growing the guiding values by which to lead one's life.

Such character building makes us more responsive to what the real needs of any situation actually are, rather than what we guess or hope they might be. And it also makes us freer and more aware and encouraging of the unique talents and contributions of other people, who will be much better able to respond to certain needs than we can.

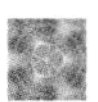

Some of the finest leaders that many recognise, are those who respond to the pre-eminent needs of a situation and time in a more selfless, reasoned and sentimented way. This enables them to really understand the needs and act upon them despite prejudice and opposition; with the knowing that however much people criticise what doesn't work, the best leadership is to put something better in its place.

To respond to a need is not to fight *against* something, but to fight *for* something, such as a principle, a cause or a vision. This nature of response was embodied by Mahatma Gandhi for example, when he rose above polarisation and factional divisions, guided as he was by the principles of 'Satyagraha' or 'holding on to truth'.

And it is the search for truth, and the living of it, which is an innate need in humans and a powerful leadership for life.

For letting the needs lead means to acquiesce to the truth of things. It means remaining unbiased, fair and determined to find and best respond to what a situation actually is and calls for, regardless of personal likes and dislikes.

Such leadership ways can guide and inspire new generations of young people, if we can live by them, whilst upholding and encouraging the best of each other.

And whilst the world still suffers from many tyrannies, the collective will of people, leading by good example, is a powerful force for leadership and change. This can be seen around the world in some spiritual, environmental and peace movements – and some revolutions too.

When all are genuinely focused on a greater need, the spaces appear for each person to find their best role or function in any venture or collective task. If this could happen more in peacetime than it does in war, perhaps there would be less warring.

Which needs lead and inspire people to greatness?

Real vision, ideas, ideals, sentiments, fine attitudes and fair principles are all compelling needs that motivate and inspire individuals and nations to greatness. Such aspirations have arisen over centuries and been characterised differently around the world, such as in France with the principles of 'Liberty, equality, fraternity', or in the United States with the 'unalienable rights' of people to 'Life, liberty and the pursuit of happiness'.

Yet what ideals, vision or principles would a nation want to rally to in these times?

Rising to ideals involves having the courage, conviction and flexibility to change and refine them as needs be. For mindsets and attitudes and high ideals need a commensurate, ongoing development in people in order to be upgraded, upheld and applied into daily living – otherwise they become merely intellectual, dogmatic, sterile or worse.

When a person lives an ideal, mindset or attitude for real, they become one of its leaders.

Yet in these times it seems that many leaders in society and government are no longer looked up to as before. Maybe this indicates that the world is looking for a leadership characterised by genuineness, forward thinking and the kind of principled guidance that these changing times are calling for. Such leadership might instinctively be recognised and respected by all, although whether the world's hierarchies, systems and vested interests would allow and encourage this is another matter.

There is a new way to catch here, to make an opening into what leadership could be now, for the future, in the light of a better understanding of the past. For history teems with great accomplishments and disasters, but more importantly, it also leaves a trail of wisdoms to help us know and do better.

Many periods in world history are full of great cruelty, and a lot of these have been researched and exposed. But the world continues to suffer and repeat these tragedies, inhumanities and genocides over generations, right up to the present day.

The release from this can begin with a leadership of responding to the needs of humanity viewed as an integrated whole mutuality; a community in common unity. A true brotherhood and sisterhood where the best of all people and nations join together in shared purpose and respect – an enduring partnership of values.

Safe progress in leadership arises from continually finding the right probing questions and the most fitting responses, whilst we learn diligently from past mistakes and move on.

For example, does history teach that a reaction from anger, bitterness and revenge ever works? Let's consider the following persistent, rallying mindset: 'War on Terror'.

Anyone who has been in a real war, with their life on the line, will tell you that war is full of terror. War is a terror. So the mindset 'War on Terror' is based on the assumption that you fight terror with terror, which has never been proven to work. It is thus illogical, ignorant. And what does it do?

Isn't the evidence, over centuries, telling us that fighting against terror mostly leads to more deaths, more terrorism and more warring in the world? And often underestimated is the profound psychological harm that all this does to children, to people of all ages; years of further killing and torture, millions of new refugees and the ongoing destruction of homes and the environment.

So a mindset of 'War on Terror', and other such reactive variations, doesn't work in principle or in practice – it just fuels more of what it presumes to cure.

And full scale wars can often be traced back to local conflicts in and between communities, when we may learn that reacting to being hurt with retaliation or revenge is valid and even desirable... and perhaps there is no one around to demonstrate and set a better example.

Reactivity rarely makes for good leadership, because you usually end up *reacting* against something you don't like, rather than *responding* to what the situation is calling for. It's easy to be very reactive to a problem, but much more challenging yet natural to take the time and care to discover what the needs are and respond accordingly.

Every great tragedy is also an important opportunity to reflect... it can become a turning moment in human history to seriously consider what kind of a generation of human beings we want to be.

A leading change in mindset for these times

Only a weak person has to prove their strength. And each person who seeks to offer leadership by trying to prove their strength, and prove that they are not weak, and prove that they mean business, in the combative rhetoric that sometimes floods our television screens, has only proven that this absolutely fails.

This leads us again to the need for a complete change of mindset, where strength is in perception, consideration and mindfulness. For in our makeup as humans it is presence of mind that leads to and naturally effects changes.

If humans are meant to be reasoned, kind, learning, improving beings, then it's qualities such as these that provide the climate for a change of mindset, for they lead to applying the power of our consciousness in constructive ways. True leadership may flow naturally from this, without the need to 'try to be a leader'.

In a more natural state of living one's life, leadership and exampleship join forces. You lead by example. This is an inner strength that you don't have to prove to anyone. It's an ongoing endeavour where, at the heart, is both a love of truth and an absence of fear about the judgements of others.

The need for change

Change is a powerful need we all face through the different seasons of our lives. And living a life which recognises and embraces change offers a reliable compass and guide to the best way ahead.

The societies we live in however tend to oppose real change because of the inherent insecurities and short-term thinking that underpin twenty-first century living. These can pressurise us to harshly compete with, fight against or demonise others to make us feel more secure. In so doing we may acquire a false sense of what it means to gain the respect we seek; in which case we are again led by self-centred reactivity rather than self-determined response. This is a state of inertia, or even going backwards.

Yet changing to meet new requirements is the most natural way to proceed. It just depends on what we are capable enough and free enough to do with the experience and intelligence that we gather.

Stop, search, re-search, take a position

To effect lasting changes in our own self-leadership, and in our exampleship towards others, there's an overriding need sometimes to come to a full stop.

If you are trying to get from Madrid to Seville and you are travelling due north, then you'll need to stop and turn around – you're going in the wrong direction. Likewise, if you find that you constantly criticise and put others down, you can't change that or any other type of set behaviour overnight.

You'll need to come to a complete stop first and ask yourself some pivotal questions, such as 'what are the importances in life?' and 'what do I want most in life?' This can provide an overview and context in which to understand why you might put others down for example, and how you would go about changing this should you want to.

In coming to a stop, in seeking to understand the truth of any situation that you are facing, you may then be able to perceive much more about the bigger picture of what is happening and why. This can direct you to what the actual needs are, which is the best place to respond from.

This can be in our daily awareness and mindfulness, in how we talk to ourselves and others about what's most important, and about how our lives can be more relevant and useful. For in so doing, we can become more versatile, better equipped and ready for change, which often arrives unannounced.

All the time, we come back to the ongoing need to be open and responsive to change. For everyone and everything is subject to change. It is the way life evolves – the way we evolve.

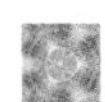

There is another, basic level of needs in these times that is pressing ever stronger: the needs that millions have right now for clean water, food, medicine, shelter, freedom, human rights and safety from oppression. But the evidence is that even if one of these issues does improve slightly over time, the situation rarely changes in spite of much aid, concern and genuine care. Is this because there remains a fundamental imbalance and lack of grasp in and between people in relation to needs and purposes beyond survival – such as the need for perception and the quality of humanity?

If this is the case, then perhaps there is an overarching need to come to a stop, and reflect with humility, insight and agreement, about what it truly means to be human and to act human. The greater comprehension and expression of this could enable us to address needs – from spiritual to survival – more from the core and causes, with a long view towards the future.

Perhaps this will guide us to the importance of *what* leads, before we consider *who*.

What are the needs that lead your life?

*

What are the greatest needs facing humanity?

A quest for life – meditations

Blend with the natural currents of life –
they know where they're going

Blend

the great intertwining of all living things

A warm breeze plays gently across the face. It carries the faint scent of sea, the warmth of sunlight, gulls crying out, children playing, feelings of longing, a subtle taste of salt, freshness, dreams. One scintilating, streaming, coursing blend...

... a tiny, vivid snapshot of the countless moving pictures in a life. One small scene, interwoven with multiple, panoramic scenes, appearing every moment in limitless vistas of intertwining, interdependent senses, thoughts, imaginings.

Such vast small worlds of endless adventures and treasures to be found in this time we have here, swept up within timeless waves of new life unfolding, pulsing back and forth in relentless tides, between and beyond the shores of greater worlds unseen and unknown.

Here we live life, and straddle the small inside the mighty.

Such infinite worlds within worlds to touch and sample, as they breathe in and out of our bodies, our minds, our senses... whilst we journey this planet realm, perhaps one day beyond, into space nearby, or distant, uncharted.

All such worlds, universes and creations, connect in a sublime meshing of one; a sacred threading of weave upon awesome weave, free-roaming within the majestic state of natural laws and the powers of evolving refinement.

Maybe, sometimes, we feel the close-in stillness of such teeming, latticeworked, filigree movement of life itself, as it urges us quietly, patiently onwards.

Such sublime feelings to be felt, now, inside it all, as a connected, tiny, working part, moving as it moves, in rhyme, in rhythm, in concert, with the graceful humming of universal notes, turning points, passages without start or end, unseen power lines, shapeless energies in countless dimensions. All an infinite same purpose of different possibilities, where each conscious effort to refine, resonates and sends its refinement to every part.

Such wondrous harmonies and patterns to life's intricate display – blended, whole, and irrevocably joined to the human cause.

We live only because of such intelligent, benign, interconnected states; these the overflowing wells of sapience, from where we derive our own signature as humans.

Such is the ineffable beauty of unforced integration, like a family who simply love to be in the endearment and company of each other as a one evolving union and kinship.

How glad can all this make us, as a basic gratitude for such profound inclusion? For such wonder never ceases, and beckons us away from the other side of the divide, where the strain of disconnected living, and its mirage, beguile us to be lesser.

Invitations from the gods arrive with every passing moment, in the new born day, the next understanding, a turning moment, any of the infinite experiences to be had in our brief stay on earth.

And with the resounding perception that all life is a blend of different ways and levels and functions, comes the certain knowing that we are never alone.

What do eyes see?

We each know a little bit about black holes: we have a small one in the centre of each eye. They let light in and can give a rare light out.

Yet we can look without seeing, or see without perceiving. So the more open-minded, the more open-eyed.

It's a wonder then what those universal black holes out there see...

Perhaps our understanding would be helped by looking at our reflection and observing our own eyes. What have we taught them to perceive? And is looking into ourselves harder than looking out through a telescope?

When we look inside ourselves, in the brilliant light of what we are made for, we may really see and feel the pulse and purpose of life. Our seeing and perceiving is our bright torch into the magnificent unexplored dark of the universe within and without.

Is this why we have these two ever-changing pupils, our personal black holes, letting impressions pass in and out, without judgement, without cynicism, without conditions? Up to us then to keenly translate what we see, and why this may be different to what others see – even when we are looking at exactly the same thing.

What an amazing phenomenon then, that in those moments when we love what we are, what we do, or who we are with, these two little black holes can dilate to let in and give out more illumined light of such experience. What brilliance created this, for us to be able to see a glimmer of what this brilliance sees?

What a sublime design of white, black and colour is the eye, where white is the mix of all colours and black is the absence of colour and an invitation to receive.

Wherein the coloured iris is the eye's unique portrait, depicting what we do and have done with our lives, within the covenant we are granted to live.

This surrounding circle of the iris – meaning rainbow – controls and enhances the light pathway of the pupil. And in a similar way, the planet's own rainbow nature and natural laws, order and liberate what a human can be and do.

The wise iridologist reads the iris of a person and, like the good palm reader, looks into the map of their life's journey to perceive and translate. For the course of our life leaves many trails and traces.

Look at the shine in our eyes when they radiate the fullness and warmth and wonder of being alive. Such eyes see beyond normal sight – through matter, distance and pre-conceptions. Eyes are indeed portals into another world.

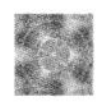

If we let our eyes open, to scan freely, far beyond their physical seeing, perhaps the story that they would tell might begin like this:

You see us,
and yet perhaps you do not,
for we are eyes made in reflection
of the great Visionary
which first saw the light
of a human
and designed their being.

And this vision had no hues,
it was a colourless dream
of partnership, love and honour
in spirit light passing as waves
across the universe
travelling with the hope of finer light
through the being of human
and outflowing in return response
to the Visionary
when the yield of human endeavour
is freely given away.

You are of the rainbow maker
of originality
as your iris sees it
save when you are blinded
by too much self.

We will serve you
even when you do not see
how graced your lives are.
For we give you vision,
further than any other living being on earth.
And when you look to the galaxies
your eyes softened, clear,
you will see that brilliant stardust
of high reasons
which gave you birthright.

Close your eyes
use your inner sight
slowly
to reflect on the ineffable beauty
of your design;
in what it can be from the black,
and do from the white.

And when you open your eyes again
you may begin to see
what we see.

Seeking the source of life

a flow of meditation starters

When does a human become a human with being?

We are born to be sensitive instruments of change and growth. A human with being is responsive and true to their sapient design and the natural pathways of growth. No need for titles, medals, acclamation, applause.

When a human is being themselves, there is a peace of mind that transcends 'I should or I ought to'; a peace of mind that listens to the quiet insistence of callings and needs as they travel through this universe and echo in sameness in the human, releasing the desire to seek and serve high causes.

This is the silence from which original ideas and music and language and creativity arise free, unrestrained and uncensored.

This is the stillness from which great endeavours are born.

This can be described as a 'colourless' inner state, in that it has no attachment to any bias, competitiveness, dogma, history. It is a state that flows, from being into doing, from colourless to colourful, from power into application; a state in which the cause, the process and the result become one throughout expression of living a life of infinite possibility.

A colourless state within a person is a deep seeking to know '*what* am I?' before '*who* am I?' It is about human purpose before human identity, human essence before human personality. It is a state beyond being and doing. It connects to the source of life.

This colourless, core state of stillness and high connection works like clean water, which you can drink, feel and sense, yet it has no taste or smell or colour. The water that makes up most of the substance of our bodies is at source colourless, which allows it to manifest in a diversity of colours, such as in our venal and arterial blood systems.

The colourless property of water allows it to freely shape itself, hold itself and be coloured by the purpose it serves – just as we can. When water serves the planet, it allows itself to be coloured to the planet's purposes as it forms into moisture, rain, oceans, forests, plants, animals, humans.

Water enhances, heals and cleans, as we can do, consciously, when we find and show our own true colours in the higher purposes we court and engage. This is a refining personal development and a lifetime's journey.

For a person to be natural to themselves, newly evolving with each passing day, this core inner state needs to be kept clean and upheld by coming to know what we won't do; if a person does not put down, oppress or harm themselves or another person, and knows that they will not do these things, then they set themselves free in what they will do.

From such a colourless state, the being and doing that we call living becomes a rich weave of however you choose to colour the way you live.

But this colour is not, as a first principle, in choosing to become a dancer, a doctor or a gardener, but in knowing the reasons why you do what you do. These reasons may either connect into the purposes and source of life, and therefore be promulgated into the future, or not.

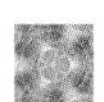

The process of being human and doing human from the improving best of oneself is honourable, for it honours what we are each uniquely capable of; it honours the source of life, and humanity, which is created from that source.

Again it happens just like the coming together of water, in which the following is a free-flow of reasoning that fits no classic form or textbook:

*Water is formed of two gases that are themselves colourless: Hydrogen and Oxygen. Hydrogen is the most ubiquitous element in the universe, and we could use its symbol '**H**' here to represent a colourless state; and 'H' can also stand for 'Human'.*

*Then there is the breath of life, which the human draws from the Oxygen on this planet, as it breathes in its nourishment to energise its systems to work and generate. We can use Oxygen's symbol '**O**' to represent this; 'O' can also represent the planet.*

So the universal Hydrogen meets and mixes with the planetary Oxygen, creating various media and states of water. There is 'water seen', for example in oceans, clouds and dew, and 'water unseen', in the air; all states support life and evolution.

*If we then add Nitrogen, which is also a colourless gas that makes up the majority of air on planet Earth and carries the symbol '**N**', we discover the trio of Hydrogen, Oxygen, Nitrogen which are amongst the most plentiful gaseous, colourless, essential constituents of life on and above earth. And as their suffixes highlight, these gases are all an integral part of the GENeration of life in this universe.*

The acronym of these three elements reveals the prefix HON, as in HONOUR.

And does whatever causes these three gases and potencies to exist and blend – as they support and promote life on earth and the very air we are breathing now – not do honour to humans and to all existence on this planet? And is honour not a main pathway that we travel when we pay tribute for the gift of a life in this universe?

This whole line of thinking, in its flow and currents, began with and has led back to the quality of honour. It may seem an unconventional trace of reasoning, but in seeking the source of what causes and creates life itself in all its mystery, conventional straight-line thinking will not work – after all, what precise straight lines exist in nature?

Mystery calls for curiosity, wonder and a free range of thinking the unthinkable, whilst following traces and clues beyond the current stock of knowledge.

So the flow continues…

… for we can engage the power and quality of honour as we grow and mature: honour from oneself to oneself in being genuine, and honour to other people, in one's respect and value and encouragement of them as fellow humans with whom we share the same time and space. And ultimately perhaps, honour from oneself to our creator, when we are not being different, offset, or perverse from how we were created to be and do.

Our honour is in making a tribute and contribution to living, as part of the human deal, within the precious grant of life.

Does this not come from a prime, core, colourless state in a human, in that it asks for nothing, offers everything and seeks no reward? And in so doing and being, it releases every colour of human expression under the sun.

Does this not then mean that we can partner and honour the very source of life itself?

If we follow the current of this flow, a question arises: where does the reason and signal come from that sparks our human spirit, our pulse, our urge to be better, if not from the source of life itself? In which case we are connected to and upheld by that source all the time, as a birthright, whether we care or not, or are conscious of it or not.

The forces of Creation send towards this end of the universe, states in which we can find and grow ourselves, granting us the supreme opportunity of making a return, a refinement, back to the source, and in furtherance of life itself. Is this not fundamental to our reason for being alive?

We are in such an open-ended state of possibility, right now, and can consciously, willingly and responsively draw from and give to the source of life all the time – except when we act as though we're the centre of everything.

a containm

nore than myself

Mortality, immortality, beyond

We are moving through life, with life. Each experience we have sculpts and shapes and reveals another facet of the meaning of our existence here on planet Earth… and sometimes of the life that beckons beyond.

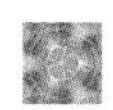

Sitting quietly, dwelling, looking out into the garden… so many natural analogies to be found here, to explain and open up the wonder behind what we see.

In the summer glow, stand the annual flowers in their resplendent hues, inspiring, lighting up, months of fragrant fireworks, and then gone.

Beside them, for years in the garden, grow the perennials, spreading, blooming and filling the air with their theatre of fine aromas and charms – an enduring sight for the times to come.

It's the same with us humans.

Our body and soul are like hardy annuals, threescore years and ten for us, more or less; growing strongly in our youth, soon to flower in mid-season; then to mature and withdraw, where in our passing, body and soul are yielded back to the planet, whence they came.

Whilst overarching this, the perennial chance at life awaits, to be seeded, cultivated, by a sapient mind and spirit, in their striving and longing and high reaching enterprise. That we may grow a vine, with tendrils stretching way above the planetary domain, into an eternity of usefulness. A life beyond the one-off annuals, however hardy.

Such a perpetual, perdurable, universal possibility emergent. No longer subject to the planet's laws and seasons. A life exalted by its high connections, made in humility and humanity, trained on earth, transformed thereafter.

Then perhaps, a state beyond such immortal trajectory; a sacred existence, with multiple seeding of new continuance, at different levels, in different dimensions, with different functions, a 'never knowing the end' possibility... somewhere beyond the far reaches of this universe?

Now, patiently reflecting, slowly turning inside, into our own complex, faculties, aura, perhaps we feel and distinguish those parts within us which are but mortal: the exquisitely versatile brain and senses and instincts and body we are given, and the beautifully willing soul that orchestrates and offers rhythm, harmony and timing to it all. What a blessing to behold and uphold.

This mortal, charged network of aptitude, powers and movement; in first bloom in the early years; supple, strong and forceful into the middle years; to cradle the release of a growing stature, prescience and momentum into the later years.

Thus the service of mortal to immortal, where the soul, brain and body support and enhance the liberation of the self-developed, conscious, immortal life from within; like the two-stage rocket, where the first stage gives fuel and propulsion to the next, is then spent and falls away, releasing the second stage far into space.

And so the immortal springs free of its mortal coil, a sacred release, like the moment a butterfly rises into air, having strengthened its wings against the sides of its decaying chrysalis.

Is this a second quickening, when our very essence and motives and inclination align through the starfields – a greater destiny to become? Is this the transcendent life, ready, as we make ourselves ready? Such immortal possibility, there to be understood, tended, on purpose. So to pass on. So to live on.

How amazing then, that we could make such a deliberate, chosen development; to refine what we are, out from this nursery school here on earth, as we play and learn to express ourselves usefully through the human form.

Immortal then to become; no competition, no exam to be passed, no transaction for personal gain, no bargaining or persuasion, nothing to keep if it cannot be given away... a high intentioned life, fed by the power of our determination and readiness to be unconditional in response to the eminent tasking to be human.

Immortal is the core and the seeds that grow of their own account in us, when we love and tend the vital sap of life that fills our very being; only wishing to serve its continuance as a first principle, not our own.

Now again, searching deeper still, perhaps there stirs a heightened care to the mortal of oneself: the good habits, foods and mind diet of daily living, the healthy exercise of one's thoughts and actions. For this, we may reason, is the soil to be tilled with respect, to be seeded with the immortal vision and aspiration and yearning of a deeper self, as it takes root and stems and flowers, to perhaps one day make a new seed of continuance far beyond this world.

The mortal of us is the great adventure of learning, the wiser to become, the fun to have, the bad times to expend, the silly things to do that won't matter later but are still to be done, the joy and peace and wonder of being born into the splendour and 'can do' of this magnificent human carriage...

... whilst the immortal awaits our awe of it, our nurture of it, our dedication to endeavour and persist and seek to understand our purpose each day a little finer – to join and partner great causes. What beauty then, that we may play some small part in it all, adding the pure oxygen of our best to such creative fires, joining an eternity, an illumination and purpose shared, a future calling, there before us.

Such a covenant and promise to living. A 'thou mayest' sacred chance at life. Endless fields of promise, where the immortal spirit of human endeavour can be freed, through our will and willingness, and by the yield of our honest application. From such a crop as this, to be harvested, in our last season on earth, with the light of our experience – if light be our tender.

You, yourself, together, when whole, are the glory and satisfaction of the most high; that humans may become, with sisters, brothers, companions of kind, the jewels in its crowning.

Such light of lights awaits our journey – inconceivable in this earthly garden – as the future flashes and shines in eyes that learn to see and feel and touch the warmth of such illumination, and what purpose it exalts and magnifies.

If immortality reaches this high, but is not an ultimate, what other unknown states and functions and realms await beyond?

Could it be that our lives are really a highest calling, to respond naturally, as we discover and live what 'naturally' means?

And is the fuller promise then, that we may become a part and partner to the calling itself, and exist and serve in a life beyond time, beyond the deepest space, beyond the comprehension of states mortal and immortal?

Do you believe that your life on earth is a starter life,
and the first stage of a universal life?

*

If so, how does this affect the way you see yourself,
and how you will spend your time today?

After all

In the end

In the end
we either believe our own exaggerations
our self-deceptions

or we nurture
the skill and courage
to search
into the heart of truth
and acquiesce to its leadership.

In the end we either live an illusion
of an identity formed by what we own
what we like or dislike

or we choose to live
in the constant rough and tumble
real unknown
of being at the point
with what really matters
to whatever sublime beauty
lets us breathe, feel, act.

In the end
we either meet an end
or we find a life of service
to that which has no end.

Lambent time come

He was scrying
through the window and beyond
his desperation
ripping at the townscape
piercing at its obduracy

In the half-light
no one was to be found
except himself
in himself
a slender flare of lambent blue
a colour code now endangered
a way to be rejoined

He noted the contrast
the flare of flame
shadowed by the tiring of faith
that predictable conduit
of the ambient monster
imposing its will
upon every vulnerable
piece of mind
sans peace

It will rise
so silent
when the culmination
finally culls
and the new void
will vibrantly unfurl
its fine nervous system
throughout this part of the universe
and in those parts
of human volunteers
left to serve
and partner
the change

It's not what you think

Don't you sometimes feel
very alone
as a struggling part
of a mass of human life
divorced from the sense
that there is any high purpose for living
or anything lasting left
to believe in?

But in an unexpected
sublime, compelling moment
millions of people around the world
were surprisingly caught, uplifted,
when a lady called Susan Boyle sang
in a beauty of resonance
and brave self-belief
that we'd almost forgotten.

And we all felt joined
to the notes of a soaring hope
in the sheer power and passion
of what one human being
can uniquely express
and share out loud.

So we applauded on our feet
this thrilling feeling inside
and the magic carpet ride delight
in joining a fine human sound
that instantly networked the world...

... yet thinking to ourselves
that we were merely being moved
to stand up and applaud a singer.

A contemplation for now

This moment
Is all I have
And inside this moment
Shall I wait
Listen
Breathe
Quietly
Deeply
And wait again
For a few minutes more
To see if I can feel
That essence
Of reverence
For what I am
And for what life is

So shall I seek
Such peace
And try to remain
As much as I can
Intact
With the love of life
For Itself
And the longing
And compulsion
To give of my best

So shall I treat myself
Well
And other beings
In kinship
In kind

The great chorus of liberating sighs

The little 'i'
Of me
My conscious self
Work-in-progress
Kindly micro world
Joins the great chorus of liberating sighs
Tonight
As I breathe in
The presence felt
Of space, conduit, opportunity
And those mighty infinite
Master-Mistress serving powers
Which in their ineffable, mysterious
Way of ways
Bestow and endow
To us humans
Their most precious of jewels
That sublime rough diamond
Of triune consciousness,
Ours to facet and magnify
To become part of the fulfilment
Of Its ever-faithful longing
To refine

Why I do this life

a declaration out from today

I do this
to try to become a human being again
and to serve this calling
always felt inside.

To try to join and work
the ways of change
forgiving myself and others
our grossly tainted legacy
and be released from
the human race's current self-view
and toxic persuasion
to not know oneself, not be oneself
to only see the world
outside out.

I do this
to free myself enough
of the impediments of these times
to feel that clean joy again
in being human,
that I may touch the best of myself
the best of other lives
and leave the least print
of disdain or misfortune.

I do this
because it is my will
as a spiritual force
to be constructive in the industry
of Creation's business –
to try fully.

I do this in the relentless urge
to make any response
that the powers and high reasons
that create humans
may find constructive.

It is a matter of honour,
of principle,
of no other option desired;
to become my own freedom
in the way natural
is meant to be,
knowing better
what I will not do.

I do this
to join the realms
of creative life
in whatever way
they see fit
and to honour my great teachers.

I do this willingly
as a living vocation,
as a destiny
to be fulfilled.

What endures?

What will live on
endure
count
when this hall falls silent
our voices to be heard no more?

For the noise of our antagonisms
and hurts and regrets
will surely expire
or cram the reaches
of the lower astral light.

What will endure beyond
is called
and expedited
into function
within the remit
of Creation's permission and purpose
a profound meeting of kind with kind.

To cast forward
our electrical vision
beyond our faces
is to gaze through
the brilliant, crystalline
vine of life
its shine encoded
with a value so deep
a state so fine
to invite our partnership
and joint crafting
of the enduring pathways
to forever.

The meshing

So let us meet
in the meshing
of all that is not surreal
and make the echoes together
of high purposes
yet to be expressed
and resonate ever more passionately
through our joy
in making the theatre
of original discovery
that the future could recognise
adding the lustre
of fineness
to the beautiful chance
that raw life has given us all
as one
whole
exalted
universal
tribe

A mindfulness

The future is unknown to me,
as is the future of my own life.

Yet when I see the fields of diamonds
that I already stand in,
the future is no longer before me
but inside me now
as a relief, a shining, a promise
beyond my feeble worries.

Can I contemplate
and deeply feel such value
trusting in that universal unknown
the light years of life
undiscovered
and develop a humility
a searching consciousness
to court the future
to partner it
express it
and be less full
of myself?

Intact

You and I
upheld
vitalised
energised
breathed into
loved
different
as one

Settle
can we
to the brilliance
of being alive
in the glory
of our human mind
and conscience.

So to enhance
the very core of ourselves
and release, forgive
absolve
the ego oppression
of ignorance
and arrogance

Settle
Let set
Let go
Let free

Settle
All is well

Glimpsing a new start hope

Twisting, uprooting forces
of prepossession
that steal your right mind
by the glitter of a better mask
and a false sense of strength
that you can be a something
with nothing better than an illusion
of beating the competition
to believe in…
this undermining world
of endless aphorisms
that tell us
'the fact is'
when clearly it isn't

In these times
we rip our souls ragged
along the shallowed sharp shores
of hollowed out lazy thoughts
that only stimulate
by becoming more perverse
and more
as we now ride the crest
of further illusion
in a world facing the recession
from any higher purpose
than self

Yet a second self awaits
to be fostered and freed
as a willing respondee
to the core premise of living
as a readiness to work
for a higher cause, a higher ideal
a higher reason
than survival

'So trust, be open, realise'
the future whispers
as it cascades and sparks
through this space and time
calling out a deep hope
and a planned appointment
with an unprecedented evolution
that the better tally of human endeavour
has kept ready
for such extraordinary times
as these